SpringerBriefs in Pharmaceutical
Science & Drug Development

For further volumes:
http://www.springer.com/series/10224

J. Rick Turner

Key Statistical Concepts in Clinical Trials for Pharma

J. Rick Turner, Ph.D.
Cardiac Safety Services, Quintiles
Emperor Boulevard 4820
Durham, NC 27703
USA
e-mail: rick.turner@quintiles.com

ISSN 1864-8118   e-ISSN 1864-8126
ISBN 978-1-4614-1661-6   e-ISBN 978-1-4614-1662-3
DOI 10.1007/978-1-4614-1662-3
Springer New York Dordrecht Heidelberg London

Library of Congress Control Number: 2011938148

Printed on acid-free paper

Springer is part of Springer Science+Business Media (www.springer.com)

Preface

The ultimate purpose of the results from a clinical trial is not to tell us precisely what happened in that trial, but, in combination with results from other trials in the drug's clinical development program, to gain insight into likely drug responses in patients who would be prescribed the drug should it be approved for marketing. The discipline of Statistics enables us to do this.

This book discusses key statistical concepts that facilitate the analysis of data collected from a group of individuals participating in a biopharmaceutical clinical trial, the estimation of their clinical significance in the general population of individuals likely to be prescribed the drug if approved, and the related decision making that occurs at both the public health level (by regulatory agencies when deciding whether or not to approve a new drug for marketing) and the individual patient level (by physicians and their patients when deciding whether or not the patient should be prescribed a drug that is on the market).

These key concepts include drug safety and efficacy, clinical significance, statistical significance, and benefit-risk estimation. All of these facilitate decision making during drug development, and also during pharmacotherapy once the drug is marketed.

Contents

Chapter 1
Setting the Scene

Abstract The ultimate purpose of the results from a clinical trial is not to tell us precisely what happened in that trial, but, in combination with results from other trials in the drug's clinical development program, to gain insight into likely drug responses in patients who would be prescribed the drug should it subsequently be approved for marketing. The discipline of Statistics, which can be meaningfully regarded as a way of conducting business during new drug development, provides the structural architecture within which all stakeholders operate, and enables the provision of compelling evidence of safety and efficacy by using procedures that everyone has agreed to honor. Not everyone needs to be an expert in Statistics, but everyone engaged in drug development can benefit considerably from a solid conceptual understanding of drug safety and efficacy, clinical significance, statistical significance, and benefit-risk assessments.

Keywords Statistical concepts · Clinical trials · Drug safety · Drug efficacy · Statistical significance · Clinical significance

1.1 Introduction

The discipline of Statistics is a central component of drug development. It is true that not everyone needs to be an expert in Statistics, but the discipline is of such fundamental importance in the decision-making processes that occur within drug development that everyone engaged in these endeavors can benefit considerably from a solid conceptual understanding of drug safety and efficacy, clinical significance, statistical significance, and benefit-risk assessments. Once you have a good understanding of these statistical concepts you will realize that everyone involved in clinical trials needs to play their role in the collection of optimum

J. R. Turner, *Key Statistical Concepts in Clinical Trials for Pharma*, SpringerBriefs in Pharmaceutical Science & Drug Development, DOI: 10.1007/978-1-4614-1662-3_1, © The Author(s) 2012

quality data in each and every trial. Optimum quality data lead to optimum quality information and optimum quality decisions throughout the drug development process. The goal of this book is to guide you to this realization while presenting the absolute minimum amount of mathematical formulas and calculations.

Statistics can be meaningfully regarded as a way of doing business. Calculations certainly need to be performed during the statistical analysis of clinical trial data, but computers are supremely placed to do that part of the overall process: The contribution of professional statisticians is much richer than simply number crunching. Statistics provides the cornerstone of designing clinical trials capable of generating optimum quality data whose analysis forms the rational basis of decisions. It provides the structural architecture within which all stakeholders operate, and enables the provision of compelling evidence by using procedures that everyone has agreed to honor.

Ethical conduct is absolutely critical at all stages of clinical research. It is appropriate to remind ourselves frequently that our work has a real impact on patients' lives. New drug development is a complicated and difficult undertaking, but one that makes an enormous difference to the health of people worldwide. It is a noble pursuit, and a privilege to be involved in this work. However, with this privilege comes the responsibility to conduct our tasks with constant due diligence and to the highest ethical standards. Derenzo and Moss [1] captured the importance of ethical considerations in all aspects of clinical studies in the following quote:

> Each study component has an ethical aspect. The ethical aspects of a clinical trial cannot be separated from the scientific objectives. Segregation of ethical issues from the full range of study design components demonstrates a flaw in understanding the fundamental nature of research involving human subjects. Compartmentalization of ethical issues is inconsistent with a well-run trial. Ethical and scientific considerations are intertwined (p. 4).

1.2 The Discipline of Statistics

In the realm of clinical research the discipline of Statistics (recognized here by the use of an upper case "S" to differentiate it from discussion of individual statistics such as the average of a group of numbers) can be thought of as an integrated discipline that is important in all of the following associated activities [2]:

- Identifying a research question that needs to be answered;
- Deciding upon the design of the clinical trial, the methodology that will be employed, and the numerical information (data) that will be collected;
- Presenting the design, methodology, and data to be collected in a Study Protocol. This study protocol specifies the manner of data collection and addresses all methodological considerations necessary to ensure the collection of optimum quality data for subsequent statistical analysis.

- Identifying the statistical techniques that will be used to describe and analyze the data in the study protocol (or an associated Statistical Analysis Plan that is written in conjunction with the study protocol);
- Describing and analyzing the data to evaluate whether there is compelling evidence that the drug is safe and effective.
- Presenting the results of a clinical study to a regulatory agency in a clinical study report and presenting the results to the clinical community in conference presentations and journal publications.

The mathematical calculations that are done while designing a clinical trial (determining the chosen sample size) and after the collection and organization of categorical and numerical representations of biological information (more commonly known simply as data) are straightforward and by far the easiest aspect of the discipline of Statistics. The more difficult aspects are the appropriate implementation of the art and science of the discipline, difficulties that argue powerfully and persuasively for the involvement of statisticians from 'alpha to omega,' i.e., throughout the entire process of conceptualizing, designing, conducting, analyzing, interpreting, and reporting a clinical trial.

It was noted in the previous section that everyone involved in conducting clinical trials can benefit from a conceptual understanding of the discipline of Statistics. The same is true for all health professionals involved in pharmacotherapy. Physicians who, in consultation with their patients, make decisions to implement a therapeutic intervention (e.g., pharmacotherapy, chemotherapy, radiotherapy, surgery) need a keen understanding of benefit-risk assessments since they directly influence the choice of intervention. They need to be able to read medical journals and other legitimate sources of empirical research findings, assess the quality of the information provided, and decide on a case-by-case basis whether the course of action of interest is likely to have a favorable benefit-risk profile for a given patient. As Katz [3] observed, "All of the art and all of the science of medicine depend on how artfully and scientifically we as practitioners reach our decisions. The art of clinical decision-making is judgment, an even more difficult concept to grapple with than evidence."

1.3 Generalizing Information Gained from a Clinical Trial

While precise knowledge of the data obtained from the subjects participating in a clinical trial is vital in making 'go/no-go' decisions within a drug's clinical development program (i.e., whether to progress to the next step in the program), the ultimate goal of conducting a series of related clinical trials is to form an educated estimate of what is likely to occur in the general population with the disease of interest should the drug be approved for marketing by one or more regulatory agencies and then prescribed to these patients.

This book uses a central example of developing a new drug to treat high blood pressure, or hypertension. Such drugs are called antihypertensives. This example has been chosen for three reasons:

- Everyone has a blood pressure, which is measured in various circumstances (e.g., visits to doctors' offices and when applying for a new life insurance policy);
- The units of measurement, millimeters of mercury (mmHg), are the same worldwide (which is not the case for other measurements such as cholesterol levels);
- High blood pressure is a clinical condition of great concern in many geographic regions.

In the United States, for example, around 33% of adults have high blood pressure, a major risk factor for heart disease, stroke, congestive heart failure, and kidney disease [4]. It is simply not possible for all of these tens of millions of individuals to participate in trials conducted during a new antihypertensive's clinical development program. Therefore, on the basis of the Phase III trials conducted at the end of the program, in which (only) several thousand subjects are likely to have received the drug, it is incumbent upon the Sponsor's statisticians and clinicians to provide their best estimate of how safe the drug would be, and what degree of benefit it would bring to the general population of hypertensives should it be approved.

Computations involving the actual blood pressure data collected in the trials must be used in the calculation of this estimation, which must then be presented in complex regulatory documents that will be scrutinized by regulatory reviewers. One question therefore is: How do Sponsors collect and then analyze and interpret data from a small subject sample in a clinical trial to enable the best possible generalization to an extremely large number of individuals? And a second question is: How do the regulatory reviewers evaluate the documents provided to them by the Sponsor and hence decide the drug's regulatory fate? The answer to both questions is that they use the discipline of Statistics.

1.4 Blood Pressure

Blood pressure measurement is a component of virtually all clinical trials. In the majority of cases it is a safety measure. Drugs for all diseases except hypertension are not supposed to affect blood pressure, and, since blood pressure is such an important parameter, we need to make sure that the drug is not changing (raising or lowering) blood pressure. In the case of new antihypertensives, where the drug is intended to lower blood pressure, it is an efficacy measure. The term efficacy refers to a drug's ability to do what it is intended to do, in this case to lower blood pressure in hypertensive individuals. Efficacy is a measure of benefit.

1.4.1 The Physiology and Measurement of Blood Pressure

There is continuous pressure in the arteries that provides the necessary driving force to propel blood through the capillaries, where oxygen is given to body tissues and carbon dioxide collected for transport back to the lungs via the venous system. Blood is ejected into the body's arteries by the heart. The level of pressure fluctuates during each cardiac cycle. Commonly cited healthy arterial blood pressure values for adults are a systolic blood pressure (SBP) value of 120 mmHg and a diastolic blood pressure (DBP) of 80 mmHg.

This unit of measurement is used because the pressure in the artery, if channeled to the bottom of a column of mercury, would cause the mercury to rise a certain number of millimeters in height. The first invasive methods of measuring blood pressure employed such a mercury column, and, given the fluctuation that occurs throughout each cardiac cycle, the height of the mercury varied throughout the cycle. While this invasive (direct) methodology is precise, it is cumbersome and carries risks associated with all invasive procedures. A modified and noninvasive version of this procedure is therefore practical. Such a procedure is used to measure blood pressure in current clinical settings each time a mercury sphygmomanometer is used in conjunction with a blood pressure cuff and a stethoscope.

The reason for the use of a mercury column is that mercury is considerably heavier than liquids such as water, which would initially appear to be a much more convenient (and non-toxic) option. Water could certainly be used theoretically, but, since it is approximately 13 times less heavy than mercury, the sphygmomanometer column would need to be 13 times as high. The height of ceilings in typical clinical settings and the difficulty of reading the level of the liquid at the heights that would be needed effectively preclude such an option.

This method of blood pressure measurement commences with the blood pressure cuff being placed around the upper arm. The cuff is then inflated to a fairly high pressure, e.g., 180 mmHg, and the stethoscope is placed over the brachial artery where it passes across the upper fold of the elbow. The goal is to inflate the cuff sufficiently so that the pressure in the cuff is greater than the pressure within the artery. If the pressure in the cuff is indeed greater than the pressure in the artery (as it usually will be at 180 mmHg) the brachial artery is temporarily collapsed, and hence no blood can flow through the artery. When no blood is flowing through the artery the stethoscope detects no noise.

The cuff pressure is then gradually reduced, and hence the pressure being applied to the brachial artery is gradually reduced. Eventually, a point is reached when blood starts to spurt through the artery with each heart beat. The sounds made by these spurts, called Korotkoff sounds after the method's founder, are detected via the stethoscope. The person performing the blood pressure measurements notes the pressure registered on the sphygmomanometer at the time of hearing the first sound. This is the value designated as SBP. With further reduction in the cuff pressure the blood flows more and more smoothly as the artery returns to normal. When the artery has returned to its normal state, blood flow is completely unimpeded and hence continuous. At this point, Korotkoff sounds are no longer heard. The pressure at which they disappear is the value designated as DBP.

A clinical trial involving an antihypertensive drug requires assessment of the drug's efficacy. This necessitates measurements at (at least) two time points: an initial measurement before treatment has commenced, often called a Baseline measurement, and a measurement at the end of the treatment period. Realistically, readings may be made throughout the length of the treatment period to allow a finer resolution examination of change over time. However, since analyzing data from multiple assessment points requires more sophisticated analytical strategies to be discussed, we will consider only the Baseline measurement and the end-of-treatment measurement.

1.4.2 A Cautionary Tale About Blood Pressure Measurement

Having our blood pressure measured is a common experience when visiting a doctor's office, since it is an important aspect of general health and often used as an initial diagnostic tool. However, a word of caution is appropriate. It is remarkably (and, arguably, scarily) easy to obtain *a* blood pressure: it is much more difficult to obtain the *correct* blood pressure reading.

Imagine the scenario of attending your doctor's office for a routine physical examination. It may be the case that you are running tight on time, and perhaps driving a little quicker than usual. You may also walk a little faster from the parking lot to the entrance to the office. Since you are a few minutes late, the nurse is actually ready to take you straight back to the examination rooms. Following a cursory measurement of your height and weight, the next item is usually a blood pressure measurement. The nurse slips a measurement cuff on your arm. Imagine that it is wintertime, and you happen to be wearing a long-sleeved shirt or blouse and a wool sweater. It is quite possible that the cuff was placed around these items of clothing, rather than on your bare arm. The nurse then quickly inflates the cuff, and then deflates it to obtain a blood pressure measurement. This single reading is written on a sheet, and that is that. Is there anything wrong with this picture?

Yes there is, and, unfortunately, quite a lot. First, rushing to the doctor's office is likely to elevate your blood pressure. Before any assessment is made it is appropriate to have a period of time during which you rest quietly. Second, placing the cuff on top of various layers of clothing is not optimal. Third, a single measure under any circumstances invites a degree of variability that can be avoided by a series of readings (perhaps triplicate) from which mean data are calculated.

In light of these discussions of how difficult it can be to get a representative blood pressure from a single measurement taken in the midst of multiple activities, it becomes apparent that considerable care must be taken in any setting—clinical trials being of primary interest here—to acquire optimum-quality data. Moreover, in clinical trials, additional acquisition of various data means that even more attention needs to be paid to blood pressure. For instance, if taking blood samples is part of the trial's procedures, this should not happen immediately before blood pressure readings are made: The act of drawing blood can certainly elevate blood pressure in many, if not all subjects.

Finally here, it should be noted that while there are two components to commonly presented blood pressure measurements, systolic pressure and diastolic pressure, from now on we will simply use the phrase 'blood pressure' and regard this as one number: this facilitates explication of the points to be made. This is a legitimate option, since the parameter of mean arterial pressure (MAP) is a single value that is influenced by both SBP and DBP, and is a way of expressing a representative pressure over the entire cardiac cycle. MAP is calculated as DBP plus one third of the pulse pressure, which is defined as the difference between SBP and DBP. That is:

$$MAP = DBP + 1/3 \ (SBP - DBP)$$

It should be noted in a statistics book that, while the name mean arterial pressure contains the word 'mean,' it is not calculated in the same manner as the 'regular' mean would be calculated, i.e., as the sum of SBP and DBP divided by the number 2. Imagine an SBP of 120 mmHg and an associated DBP of 80 mmHg. The regular numerical mean value for 120 and 80 would be 100. However, when calculating the MAP for an SBP of 120 mmHg and an associated DBP of 80 mmHg, MAP is 93.3 mmHg. It is a weighted average that reflects the fact that the pressure in the arterial system is around the DBP level for approximately twice as long as it is around the SBP level for each cardiac cycle.

1.5 Statistical Concepts and Nomenclature

It was noted in the book's Abstract that the concepts of current interest include drug safety and efficacy, statistical significance, clinical significance, and benefit-risk estimation, since each of these facilitates decision making during new drug development. These are introduced at this point.

1.5.1 Drug Safety

No drug is immune from the possibility of causing adverse reactions in certain individuals who are genetically and/or environmentally susceptible. The concept of drug safety must therefore be considered, and hence operationally defined. One relatively simple but nonetheless meaningful conceptualization of safety is as the inverse of harm: the less a drug's toxicity, the greater its safety [5]. What is measured and recorded can be better described in actuality as drug harm. However, in practical terms, two considerations that would argue against this term come quickly to mind. First, it may seem strange to some to define something (safety) in terms of the absence of something else (harm). Second, while drug safety has legitimately attracted considerable attention among many stakeholders, it has accordingly become a high profile topic in many mass communication media. In this context, the term drug harm would likely be incendiary, attracting even more emotive behavior and sensationalism to an area already fueled by far too much of both. The critically important assessment, discussion, and communication of drug safety issues must be approached in a calm, scientific, and clinical manner.

A useful definition of drug safety was provided by the FDA's Sentinel Initiative ([6]; see also [7]):

> Although marketed medical products are required by federal law to be safe for their intended use, *safety does not mean zero risk*. A safe product is one that has acceptable risks, given the magnitude of benefit expected in a specific population and within the context of alternatives available.

This approach will be discussed in more detail when addressing benefit-risk estimation in Chap. 6.

1.5.2 Drug Efficacy

The term efficacy refers to a drug's ability to do what it is intended to do. For example, a drug for hypertension needs to lower blood pressure to demonstrate efficacy. The more the blood pressure is lowered, the greater the degree of efficacy. Efficacy is therefore a measure of the drug's beneficial therapeutic action with regard to the disease or condition of clinical concern for which the drug is being developed. It is a measure of benefit.

Evaluations of efficacy are couched in terms of statistical significance and also clinical significance.

1.5.3 Statistical Significance and Clinical Significance

Statistical significance addresses the reliability of the treatment effect: If approved for marketing and given to a large number of patients, is there compelling evidence that it will indeed lower blood pressure? It is certainly possible that the results of a given trial could achieve statistical significance but not clinical significance. That is, compelling evidence could be provided that the drug would reliably lower blood pressure in the general population of hypertensives, but only lower it by a small amount, an amount deemed not to be clinically significant.

One kind of clinical trial, called a superiority trial, is designed to look for evidence that a new drug under development is superior to a control treatment. This control can be a placebo—a tablet that is manufactured to be identical in size, shape, color, taste, and smell to the tablet containing the drug but which has no pharmacologically active ingredients—or an active control, i.e., another drug. Throughout this book the example of a Phase III clinical trial of a new drug to lower blood pressure is used to illustrate many statistical concepts. At the end of the trial the mean response of all the subjects in the trial who received the drug will be compared to the mean response of all the subjects who received the control drug, and for the sake of simplicity this will almost always be a placebo in our examples. Although the placebo has no pharmacological activity, it is not unusual for subjects receiving this treatment to show a decrease in blood pressure by the end of the trial. Imagine that the mean decrease in blood pressure for those receiving the drug is 4 mmHg and the mean decrease in blood pressure for those receiving the placebo is 2 mmHg. Clearly, there is a greater decrease associated with the drug than with the placebo: the drug lowered blood pressure 2 mmHg more than the placebo. The question of interest, however, is this: Is there a statistically significantly greater decrease in blood pressure associated with the drug than with the placebo?

As a path to understanding the concept of statistical significance, ask yourself the following questions. How much credence would you assign to a difference of "only" 2 mmHg? Expressed another way, would you feel that this result provided compelling evidence that the drug truly is 'better' at lowering blood pressure than the placebo, i.e., that it works? You may feel that this difference of 2 mmHg between the groups is so small that it is not real, and hence believe that this result could have occurred by chance alone on this occasion. On the other hand, you may be convinced by this result that the drug really does work. The next question is: Whatever your thoughts, if you asked the same question to ten of your colleagues, how many do you think would agree with you? All of them? Half of them?

The point here is that there is no standard answer to these questions, since anyone's answer will be subjective. In contrast, the discipline of Statistics facilitates the removal of subjectivity. The scientific, clinical, and regulatory communities have all agreed in advance to accept that, for any given set of data, statistical testing provides a precise statistical answer that has effectively been agreed upon as objective. The concept of statistical significance facilitates this objectivity by expressing the outcome of a test in a way that everyone has agreed to honor.

The appropriate statistical test will provide one of two answers. One possible answer says that the group means differ statistically significantly, allowing the statement to be made that this difference is unlikely to be the result of chance alone, and, in fundamental terms, the drug works reliably. The second possible answer says that the group means do not differ statistically significantly; this answer can be given even though the group means will almost certainly differ somewhat. This answer allows the statement to be made that this difference could well have arisen by chance alone, and, again in fundamental terms, we do not have sufficient evidence to say that the drug works reliably. Compelling evidence is provided when the result of a statistical test is statistically significant [2].

References

1. Derenzo E, Moss J (2006) Writing clinical research protocols: ethical considerations. Elsevier, San Diego
2. Turner JR (2010) New drug development: an introduction to clinical trials, 2nd edn. Springer, New York
3. Katz DL (2001) Clinical epidemiology & evidence-based medicine: fundamental principles of clinical reasoning & research. Sage Publications, Thousand Oaks
4. CDC Home (2011) Web site of the centers for disease control and prevention, http://www.cdc.gov/bloodpressure/facts.htm Accessed 5th May 2011
5. Durham TA, Turner JR (2008) Introduction to statistics in pharmaceutical clinical trials. Pharmaceutical Press, London
6. FDA (2008) The sentinel initiative: national strategy for monitoring medical product safety. http://www.fda.gov/downloads/Safety/FDAsSentinelInitiative/UCM124701.pdf Accessed 12th May 2011
7. FDA (2010) The sentinel initiative. Access to electronic healthcare date for more than 25 million lives: achieving FDAAA Section 905 goal one. http://www.fda.gov/downloads/Safety/FDAsSentinelInitiative/UCM233360.pdf Accessed 12th May 2011

Chapter 2
Analyzing Safety Data

Abstract The safety of a drug is assessed at all stages of its life cycle, from drug discovery through nonclinical development, preapproval clinical development, and all the time it is subsequently on the market. The focus of this chapter is safety assessment during later-stage clinical trials. When a drug is granted marketing approval by a regulatory agency and therefore becomes available to prescribing physicians and their patients, the best available information upon which the doctor and patient can form answers to questions concerning the drug's safety profile is the safety information gathered during clinical trials. This information is provided to the clinician and all patients receiving the drug in the drug's package insert, or label. The content of the label will have been agreed upon by the Sponsor and the regulatory agency granting marketing approval, and it summarizes the best available data about the drug's safety at that point in time.

Keywords Drug safety · Adverse events · Adverse drug reaction · Exclusion of unacceptable risk · Antidiabetic drugs · Risk ratio

2.1 Introduction

The safety of a drug is addressed at all stages in its life history. This starts long before the drug is administered to humans during clinical trials. By the time Phase I trials commence the Sponsor will have collected considerable amounts of safety data during the nonclinical development program. Safety will be assessed throughout the clinical development program, and, if the drug is approved for marketing, it will be assessed throughout the time the drug is on the market. As in other chapters, the focus here is on assessments during Phase III trials.

J. R. Turner, *Key Statistical Concepts in Clinical Trials for Pharma*,
SpringerBriefs in Pharmaceutical Science & Drug Development,
DOI: 10.1007/978-1-4614-1662-3_2, © The Author(s) 2012

2.2 Providing Safety Data to Prescribing Physicians and Patients

Jumping ahead to the point where a new drug has been approved, consider a scenario in which a prescribing physician is discussing with a patient whether the new drug is a good treatment option (therapeutic decisions are ideally made by a 'health team' comprising the physician and the patient). What kinds of information might the doctor and the patient find useful? Examples include [1]:

- How likely is the patient to experience an adverse drug reaction? (The term adverse drug reaction is employed once the drug is on the market. Before that, as we will see shortly, the term adverse event is used in clinical trials).
- Are the typical adverse drug reactions temporary or permanent in nature?
- How likely is the patient to suffer an adverse drug reaction that is extremely serious, or even life-threatening?
- How might the risk of an adverse drug reaction vary with different doses of the drug?
- How might the risk of an adverse drug reaction change with increasing length of treatment?
- Are there any specific clinical parameters that should be monitored more closely than usual in patients receiving this drug?

When the drug is marketed and therefore becomes available to prescribing clinicians, the best available information upon which a clinician and patient can form answers to such questions is the safety information gathered during clinical trials. This information is provided to the clinician, and all patients receiving the drug, in the drug's package insert or label. The content of the label will have been discussed by the Sponsor and the regulatory agency granting marketing approval, and it summarizes the best available data about the drug's safety at that point in time.

2.3 The Clinical Study Report

Clinical trial data, both safety and efficacy, are typically presented in a clinical study report (CSR) that is suitable for submission to a regulatory agency. Describing, or summarizing, the tremendous amount of data that are collected in a clinical trial is typically a useful first step in reporting the results of the trial. Simple descriptors such as the total number of participants in the trial, the numbers that received the drug treatment and the placebo treatment, respectively, information concerning sex and ethnicity, and the average age of the participants in each treatment group help to set the scene for more detailed reporting. This information can be usefully summarized in in-text tables that are placed in the body of the text in CSRs. In each case, the source of the information presented will need to be cited. The source is typically one of many listings that are appended to

Table 2.1 Subject accountability (Clinical Trial ABC789)

	Number (%) of subjects	
	Drug ($N = 400$)	Placebo ($N = 400$)
Completion status		
Completed study	320 (80)	360 (90)
Withdrew prematurely	80 (20)	20 (10)
Premature withdrawals		
Adverse event	40 (10)	12 (3)
Withdrew consent	20 (5)	8 (2)
Protocol violation	20 (5)	16 (4)
Other	0	2 (1)

Other: 1. [Description]. 2. [Description] *Source Table*: XYZ

the report. Listings are comprehensive lists that provide all information collected during the trial.

2.4 Subject Demographics and Accountability

Each CSR has various sections, including subject demographics and account-ability, safety data, and efficacy sections. An in-text table may be presented for demographic characteristics. Specific characteristics that are important can vary from study to study, but typical ones include age, sex, ethnicity, and baseline data of relevance, e.g., weight, blood pressure, and heart rate. Information concerning the use of concomitant or concurrent medications and evaluations of subject adherence or compliance with the trial's treatment schedule is also typically presented.

Table 2.1 provides an example of an in-text table from a hypothetical clinical trial that summarizes subject accountability.

Several comments about these hypothetical data are appropriate. First, it is possible but unlikely that the number of subjects for the two treatment groups would be identical in a real study. Presenting percentages as well as absolute numbers is therefore useful, since the percentages allow for differing totals of subjects in each group. Second, the number of subjects in the individual categories must add up to the respective group totals. Third, explanation of (any) other reasons for premature withdrawal should be presented, either in text form above or below the table or in footnote form immediately underneath the table.

Documentation of premature withdrawals from a study is important for various reasons. The implications of premature withdrawals are different in the analysis of safety and in the analysis of efficacy. From a safety perspective, these data relate to tolerability of the drug. From an efficacy perspective, dropouts lead to missing data, and the way(s) that missing data are addressed is important from the point of view of full interpretation of the analysis presented.

2.5 Safety Parameters Measured

Safety data sets comprise enormous numbers of parameters. These include laboratory tests (e.g., liver function tests such as albumin, bilirubin, and alanine amino-
transferase), electrocardiogram (ECG) waveforms, vital signs (e.g., blood pressure,
heart rate, and weight), and possibly imaging data. These data are typically presented
descriptively. The mean, minimum, and maximum values may be presented in many
cases, which would capture measures of both central tendency (the mean) and dis-
persion (the range in this case). There may also be a count of 'values higher than the
normal range' and 'values lower than the normal range' for a parameter.

2.6 Adverse Events

General safety assessments are wide-ranging, and are typically also presented
descriptively. Safety-related data can be considered at four levels: extent of
exposure to the drug; adverse events; laboratory tests; and vital signs. We will
focus on adverse events.

During the course of a clinical trial it is likely that many subjects will have
some form of adverse events (AEs). The longer the study and the sicker the
subjects are, the more AEs there will be. When reporting the results from the
clinical trial, therefore, it is of interest to know about the frequency of all adverse
events. Several different kinds of adverse events can be distinguished:

- Pretreatment adverse events;
- Treatment adverse events;
- Drug-related adverse events;
- Serious adverse events (SAEs);
- Adverse events of special interest;
- Adverse events leading to withdrawal from the study.

There are various ways to provide this information, including listings and
in-text summary tables. Descriptive statistics for AEs typically include rates
of occurrence of the events in exposed groups overall and among subgroups of
subjects (e.g., according to age and sex) to look for any potential patterns of
differential rates of adverse events.

Often, a sponsor will report adverse events that occurred in at least a given
percentage of the subjects in either group (information for both treatment groups is
presented in each case). The meaning of the term most common must be defined
every time it is used. In this example it is defined by the statement "Greater or
equal to 10%." It is likely that the incidence of AEs will not be identical in the two
treatment groups. Some AEs that occur in 10% or more of the subjects in the drug
treatment group may occur in less than 10% of subjects in the placebo treatment
group, and vice versa. Data for both treatment groups employed will be provided

for any AE listed for either group. Data are often presented in descending order of occurrence. Similar presentations of these data are included in package inserts for marketed products.

2.7 From Descriptive Statistics to Inferential Statistics

The analysis of general safety data collected in Phase III trials is not particularly rigorously defined, and as a result the presentation of safety data, as noted, is largely descriptive. This situation is now changing, certainly for more serious events. Examples are therefore provided from the field of cardiovascular safety in which inferential statistics plays a central role.

As noted in the book's Abstract, the ultimate purpose of the results from a clinical trial is not to tell us precisely what happened in that trial, but to gain insight into likely drug responses in patients who would be prescribed the drug should it be approved for marketing. One branch of the discipline of Statistics, inferential statistics, enables us to do this. The precise data collected from the specific group of subjects who participated in a given trial are used to infer what the likely responses would be in the general population of patients with the disease or clinical condition of concern.

2.8 Prospective Exclusion of Unacceptable Risk

The prospective exclusion of unacceptable cardiac and cardiovascular risk has become a critical component of drug safety evaluation in new drug development. A useful approach to prospectively excluding such risk can be conceptualized in a three-component model comprising clinical, regulatory, and statistical science [2]:

- *Clinical science*: Clinical judgments concerning absolute and relative risks are needed.
- *Regulatory science*: Based on clinical evidence, regulatory agencies have to choose the thresholds of regulatory interest, which are, at least from an absolutist point of view, the criteria of demarcation between acceptable risk and unacceptable risk. (In practice, regulators are given more latitude based on other aspects of the drug and the disease it is intended to treat).
- *Statistical science*: Once a threshold of regulatory concern has been established by a regulatory agency (or one proposed by ICH has been adopted), accepted statistical methodologies for determining whether or not these thresholds have been breached is required. Confidence intervals play a central role in this methodology.

The statistical science component is straightforward and a precise answer to the research question is provided every time, although interpretation of the results can

Fig. 2.1 A schematic representation of the QT Interval on the ECG, and QT interval prolongation

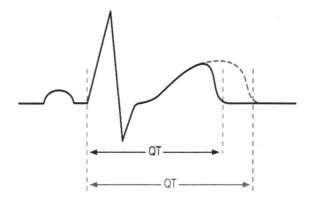

be less black-and-white. However, there is less certainty involved in the choice of a regulatory threshold. Certainly, clinical science and clinical judgment can be taken into account, but it is still neither a straightforward nor an easy decision for a regulatory agency to make. Nonetheless, for this risk exclusion model to be implemented, a decision must be made.

2.8.1 Assessment of Unacceptable Cardiac Risk

Assessment of the proarrhythmic liability of noncardiac drugs, i.e., drugs not intended for cardiac indications, has assumed a central importance in the development of noncardiac drugs. While not perfect, QT interval prolongation as seen on the ECG is the central parameter of interest in cardiac safety assessment. Figure 2.1 provides a schematic representation of the normal QT interval and a prolonged interval.

The cardiac arrhythmia Torsades de Pointes, which is very rare, often self-correcting, but potentially fatal, is associated with inherited QT prolongation, and also with drug-induced QT prolongation.

The ICH E14 Guideline addressing this issue was released in 2005 [3]. It describes the Thorough QT/QTc (TQT) study, a clinical trial devoted to the rigorous assessment of an investigational drug's QT/QTc liability, i.e., its propensity to prolong the QT interval. (Since QT varies with heart rate independently of any drug-induced effect, the measured QT interval is "corrected" for heart rate by one or more of several correction formulas, leading to the term QTc.)

TQT studies are typically conducted using healthy subjects. An indication of typical QT intervals is given by ranges of 350–460 ms for men and 360–470 ms for women [4]. The TQT study is designed to look for drug-induced increases of "around 5 ms" [3]. Operationally, this degree of prolongation is defined by placing confidence intervals (CIs) around the observed mean prolongation, in this context called the treatment effect point estimate. In this case, two-sided 90% CIs are placed around the treatment effect point estimate, and attention falls on the upper limit of the confidence interval, which is deemed the "threshold of regulatory concern" [3]. A threshold of regulatory concern is the product of the interaction between clinical and regulatory science. In this case, it is operationalized as an upper limit of a two-sided 90% CI placed around the treatment effect point estimate (point estimate of cardiac risk) obtained from a single study of 10 ms or greater.

A drug's treatment effect can be defined as the mean response due to the drug minus the mean response due to a placebo: In many experimental circumstances, although the placebo is not pharmacologically active, the mean response to its administration will still be a small change in the biological parameter of interest, here the QT interval. The treatment effect point estimate provides precise information about the degree of QT interval prolongation seen in the single study of interest. However, we wish to use all of the data collected in the study to infer what might be the responses of patients prescribed the drug should it be approved for marketing. Placement of confidence intervals around the treatment effect point estimate facilitates this inference.

Consider a scenario in which the treatment effect point estimate for QT interval prolongation in a TQT trial was 8.00 ms, and the lower and upper limits of the two-sided 90% CI placed around this point estimate are 6.50 ms and 9.50 ms, respectively. This result can be written as 8.00 (6.50, 9.50). We can now make a statement concerning the true but unknown treatment effect in the general population from which the sample that participated in the trial was drawn:

- The data from this single TQT study are compatible with a treatment effect (prolongation of the QT interval) as small as 6.50 ms and as large as 9.50 ms in the general population, and our best estimate is a treatment effect of 8.00 ms.

Recalling that the regulatory threshold of concern for QT interval prolongation is an upper limit of 10 ms, it can be seen that the value of 9.50 ms falls below this threshold. Therefore, this drug would not be deemed to be associated with unacceptable cardiac risk.

2.8.2 Assessment of Unacceptable Cardiovascular Risk

In December 2008 the FDA issued a guidance entitled *Diabetes Mellitus—Evaluating Cardiovascular Risk in New Antidiabetic Drugs to Treat Type 2 Diabetes* [5]. This guidance detailed additional research related to cardiovascular safety that sponsors must complete before submitting an NDA for a new antidiabetic drug for type 2 diabetes mellitus (T2DM). A draft Guidance containing similar cardiovascular safety requirements was released by the European Medicines Agency in January 2010 [6].

Clinical development programs for new drugs for T2DM have traditionally included small and relatively short (12-week) therapeutic exploratory studies and a

12–24-month therapeutic confirmatory study. Subjects recruited into the trials were typically relatively healthy, certainly healthier than the eventual target patient population for the approved drug. The results of all trials conducted were then presented to a regulatory agency. However, the guidances have considerably changed the nature of the required trials. A typical clinical development program will now include:

- Small, short, dose-finding early trials;
- Larger and longer late therapeutic exploratory trials;
- Larger and longer therapeutic confirmatory trials that include subjects at high risk for cardiovascular events.

The FDA guidance recommends collecting clinical trial data that are more directly applicable to the patients likely to be prescribed the drug if marketing approval is given. One consideration for achieving this goal is to make the trial protocol's inclusion criteria less restrictive, thereby opening enrolment to all patient populations that would be reflected in the approved drug's label. A second consideration is the inclusion of high-risk subjects. The guidance states that therapeutic exploratory and confirmatory trials should include subjects "at higher risk of cardiovascular events, such as patients with relatively advanced disease, elderly patients, and patients with some degree of renal impairment" [5].

In contrast to the employment of QT interval prolongation as a cardiac safety biomarker, clinical endpoints are of interest when addressing cardiovascular safety for T2DM drugs. The Major Adverse Cardiovascular Events (MACE) composite endpoint includes non-fatal myocardial infarction, non-fatal stroke, and cardiovascular death. An expanded MACE endpoint might include hospitalization for unstable angina and urgent percutaneous coronary intervention/coronary artery bypass graft surgery. Other possible secondary endpoints include carotid revascularization and lower extremity amputations/revascularization. In this context, the traditional MACE endpoint is acceptable to the FDA.

Cardiovascular events reported during the trial must be centrally adjudicated by independent experts [7]. Having all cardiovascular events for all subjects in the trials at all investigational sites evaluated against precisely the same criteria by the same group of independent experts considerably enhances the validity of the data. The guidance notes that sponsors should establish "an independent cardiovascular endpoints committee to prospectively adjudicate, in a blinded fashion, cardiovascular events" during all Phase II and Phase III trials [5]. These events should include cardiovascular mortality, myocardial infarction, and stroke, and can include hospitalization for acute coronary syndrome, urgent revascularization procedures, and possibly other endpoints.

Like the ICH Guideline E14 discussed in the previous section, the guidance takes the approach of prospectively excluding unacceptable risk by the employment of a threshold of regulatory concern and associated statistical methodology employing confidence intervals. However, the analysis to be conducted does not use data from a single trial. Rather, data from essentially all Phase II and Phase III trials are used via employment of a meta-analysis (this topic is covered in more detail in Chap. 5). It is of regulatory interest to determine to what extent (if any) the drug may increase the occurrence of the cardiovascular safety endpoints in the patient population who would be prescribed the drug if approved. In this scenario, the clinical endpoints of interest are categorical, and typically dichotomous: for example, either a myocardial infarction occurred, or it did not. In this setting, relative risk is of interest, which can be captured by a risk ratio.

Consider a hypothetical single trial in which the drug treatment group receives a new drug for T2DM, and another group of subjects receives a control treatment. The question of interest here is: Does the drug lead to an unacceptable increased risk of cardiovascular events? To answer this question, a (safety) treatment effect will be calculated in the form of a risk ratio (the term treatment effect can refer to efficacy treatment effects and safety treatment effects). A relative risk ratio point estimate is calculated by considering the number of cardiovascular events in the drug group as the numerator and the number of cardiovascular events in the control group as the denominator. If the number of events in each treatment group happened to be precisely the same (the likelihood of this is vanishingly small) the value of the point estimate would be unity, represented here as 1.00. If the number of events in the drug treatment group is greater than the number for the control treatment group, the value will be greater than 1.00.

Going one step further, since multiple Phase II and Phase III trials will have been completed, a meta-analysis can be conducted, producing a relative risk point estimate that is based on data from all of the studies rather than from just a single study. Having done this, a confidence interval can be placed around this relative risk point estimate. In this scenario, a two-sided 95% CI is employed. This information will then be used to infer what the true but unknown population relative risk is. The logic for conducting a meta-analysis first is that the inference based on a larger database (the data from all Phase II and Phase III trials combined) will be more informative than an inference based on any single trial.

Imagine that the value of the relative risk ratio point estimate computed from the meta-analysis is 1.20, and that the lower and upper limits of the confidence interval are 1.15 and 1.26. This can be written as 1.20 (1.15, 1.26). These confidence intervals allow us to estimate the true but unknown relative risk for the general patient population. We can now make the following statement:

- The data obtained from this meta-analysis are compatible with as little as a 15% increase and as much as a 26% increase in the risk of cardiovascular events in the general population, and our best estimate is an increase of 20%.

Of course, any increase in risk is not ideal. However, the question here is: Is there compelling evidence of an unacceptable increase in risk, given the serious nature of the disease? (The more serious the disease being treated, the more risk one may be prepared to take in order to obtain the desired therapeutic benefit). To permit this question to be answered requires us to compare the result with the thresholds of regulatory concern presented in the FDA guidance (such thresholds are not explicitly stated in the EMA draft guidance). Three scenarios are described:

- If the upper limit of the two-sided 95% CI is equal to or greater than 1.80, the drug would be deemed to have an unacceptable risk. In this case, the Sponsor would be expected to conduct more research and collect more data whose analysis would yield an upper limit of less than 1.80 before requesting marketing approval.
- If the upper bound is equal to or greater than 1.30 and also less than 1.80, and the overall benefit-risk analysis presented at submission supports marketing approval (benefit-risk is considered in more detail in Chap. 4), "a postmarketing trial generally will be necessary to definitively show that the upper bound of the two-sided 95 percent confidence interval for the estimated risk ratio is less than 1.3."
- If the upper limit is less than 1.30 and the overall benefit-risk analysis presented at submission supports marketing approval, "a postmarketing cardiovascular trial generally may not be necessary."

Returning to our hypothetical result from the meta-analysis conducted for the drug's Phase II and Phase III trials, the results were a relative risk of 1.20 (1.15, 1.26). As in the case of the TQT study and its investigation of QT interval prolongation, interest focuses on the upper limit of this two-sided confidence limit. This value, i.e., 1.26, falls below the regulatory threshold of concern of 1.80, and, moreover, also falls below the more stringent threshold of concern of 1.30. This result therefore falls in the territory covered by the third bullet point in the previous list. It is likely that, given a favorable overall benefit-risk profile, the drug will be approved and there will not be a requirement to conduct a postmarketing trial to further evaluate the relative risk.

2.8.3 A More Realistic Scenario and its Unintended Consequences

The relative risk point estimate and its associated two-sided 95% CI in the example in the previous section were deliberately chosen to illustrate the scenario where the cardiovascular risk associated with the drug was not deemed to be unacceptable, and was not therefore prospectively excluded by denying the drug marketing approval. However, this example represents the 'ideal case.' A more realistic example would be a case where the upper limit falls below the 1.80 threshold of regulatory concern, but above the 1.30 threshold. It is therefore likely that the Sponsor would be asked to conduct a postmarketing trial to show definitively that, based on a much larger sample of data, the upper limit actually falls below 1.30.

The intent of the guidance is an extremely laudable one: It is to ensure that new antidiabetic drugs do not unacceptably increase the risk for cardiovascular events in patients would receive a new drug should it be approved for marketing. The importance of this intent is underscored by observations that diabetes increases the risk of heart disease and stroke and that the majority of patients with T2DM die from cardiovascular disease and not from their hyperglycemia *per se*. However, there may be unintended consequences. A postmarketing cardiovascular trial of the kind discussed in the guidance requires a large number of subjects (perhaps in the order of five times as many as participated in an earlier Phase III trial), and such trials cost a considerable amount of money (tens to hundreds of millions of dollars). In addition to the costs of the 'regular' Phase II and Phase III trials, this represents a financial challenge to many sponsors. It is true that the FDA should not be concerned with such costs: their job is to protect and promote public health. However, an unfortunate and unintended consequence may be that fewer drugs for this serious disease will be developed by Sponsors who are concerned that trying to develop a drug for this indication may be beyond their financial resources.

Caveney and Turner [8] noted that an inspection of relevant data on www.clinicaltrials.gov reveals a general increase in the number of diabetes trials between 2005 and 2008, followed by a leveling off in 2009. Additionally, late 2008 and 2009 saw several smaller biotechnology companies abandon their diabetes programs because of the cost increase. It is therefore possible to speculate that the new regulatory mandates (or burdens) are leading all biopharmaceutical companies, regardless of size, to re-examine their diabetes pipelines and re-forecast their predicted return on the investment necessary to bring a drug to market. As these authors commented, "With so many patients suffering with diabetes, the unclear pathogenesis of the disease, and patients not meeting professional goals for optimal care, the field is ripe for more research discoveries and the market is open for further drug developments. Regulators, policy-makers, and industry leaders will need to be vigilant and work together to ensure that the new regulatory guidance does not stifle the development of antidiabetic agents" [8].

References

1. Durham TA, Turner JR (2008) Introduction to statistics in pharmaceutical clinical trials. Pharmaceutical Press, London
2. Turner JR (2010) Integrated cardiovascular safety: employing a three-component risk exclusion model in the assessment of investigational drugs. Appl Clin Trials June issue:76–79
3. ICH Guidance E14, The clinical evaluation of QT/QTc interval prolongation and proarrhythmic potential for non-antiarrhythmic drugs. www.ich.org
4. Turner JR, Satin LZ, Callahan TS, Litwin JS (2010) The science of cardiac safety. Appl Clin Trials 19(Suppl 11):4–8, 14
5. FDA (2008) Guidance for industry. Diabetes mellitus—evaluating cardiovascular risk in new antidiabetic therapies to treat Type 2 diabetes. www.fda.gov
6. EMA (2010) Guideline on clinical investigation of medicinal products in the treatment of diabetes mellitus (Draft). www.ema.europa.eu
7. Turner JR, Somaratne R, Cabell CH, Tyner CA (2011) Centralized endpoint adjudication in cardiovascular outcomes studies: composite endpoints, risk ratios, and clinical endpoint committees. J Clin Stud 3:46–49
8. Caveney E, Turner JR (2010) Regulatory landscapes for future antidiabetic drug development (Part I): FDA guidance on assessment of cardiovascular risks. J Clin Stud January issue:34–36

Chapter 3
Assessing Efficacy Data

Abstract A drug must demonstrate efficacy to be given marketing approval. That is, it must provide therapeutic benefit to patients with the disease or condition of clinical concern. Efficacy is assessed in two ways: there must be compelling evidence of both statistically significant efficacy and clinically significant efficacy. The former is addressed via formalized hypothesis testing involving the creation of a research question, a research hypothesis, and the null hypothesis. The latter is addressed via the placement of confidence intervals around the drug's treatment effect. While important, the demonstration of statistically significant efficacy alone does not equate to the demonstration of a biologically and medically important degree of efficacy. Statistical significance addresses the reliability of the treatment effect, while clinical significance addresses the magnitude of the treatment effect, i.e., its degree of therapeutic benefit.

Keywords Efficacy · Statistical significance · Clinical significance · Inferential statistics · Null hypothesis

3.1 Introduction

The concepts of statistical significance and clinical significance are fundamental to the assessments of a drug's efficacy. For regulators to accept that a drug is effective, compelling evidence of both needs to be provided. The discipline of Statistics enables provision of compelling evidence of statistically significant efficacy by using procedures that everyone has agreed to honor: If the results of a trial attain statistical significance, all stakeholders have previously agreed that such results provide compelling evidence of statistical significance. Providing

J. R. Turner, *Key Statistical Concepts in Clinical Trials for Pharma,*
SpringerBriefs in Pharmaceutical Science & Drug Development,
DOI: 10.1007/978-1-4614-1662-3_3, © The Author(s) 2012

compelling evidence of clinical significance is a less formulaic, in that it requires the employment of statistical methodology in conjunction with clinical judgment. Nonetheless, it is equally as important as statistical significance.

3.2 Probability

Probability is an important component of Statistics, and one that is central to providing compelling evidence of efficacy. One commonly used level of probability is the 5% level, a percentage version of odds of 1 in 20: If the odds of something occurring are 1 in 20, there is a 5% chance that it will occur. Imagine a scenario discussed by Turner [1]. Twenty playing cards are lying upside down in a row on a table. You are told that one of them is the 'Ace of Hearts.' You are then asked to select a single card and turn it over. What is the probability that the card you choose will be the Ace of Hearts? There are 20 cards, and each card has the same chance of being chosen as the others. There is only one Ace of Hearts, and so your odds are 1 in 20, and hence you have a 1 in 20 chance of picking the card. In other words, there is a 5% probability that you will choose the Ace of Hearts.

A probability of 5% can be expressed as $p = 0.05$. The statement $p < 0.05$ (i.e., a probability that is <0.05) means that the odds of something occurring is <1 in 20, i.e., <5%. It has become statistical convention that, if the p-value that results from an inferential test is <5%, the result is declared to be statistically significant.

The phrase "$p < 0.05$" may well be the single most well-known term in drug development. As explained shortly, there must be statistically significant evidence of a drug's efficacy for a regulatory agency to approve it for marketing (there must also be compelling evidence of clinical significance, as discussed in due course). However, despite its attained prominence, and similarly to the thresholds of regulatory concern discussed in the previous chapter, the value of 0.05 was not ordained: it was conceived by the visionary statistician Sir Ronald Fisher. He could have chosen another value. Had he decided, for example, that odds of 1 in 25 were more appropriate than odds of 1 in 20 in this context, the associated p-value would have been 0.04 instead. Whether the value of 0.05 is "right" (whatever right means) is not the issue here. The important point is the acknowledgment that a particular value has been chosen and honored to allow the discipline of Statistics to be beneficially employed: that value is 0.05.

From the present perspective, a statistically significant result is regarded as a probabilistic statement that the result obtained was not a chance occurrence, but was the result of a systematic influence on the data collected from the subjects in the two treatment groups. What systematic influence(s) could have been at work here? Consider a randomized, placebo-controlled, parallel-groups Phase III clinical trial investigating a new drug for hypertension. The process of randomization was used to ensure that, as far as possible, the subjects in the drug treatment group were similar to those in the placebo treatment group. Hence, the only systematic influence was the treatment administered: the new antihypertensive drug under development was administered to all subjects in the drug treatment group, and the placebo was administered to all subjects in the placebo treatment group. By statistical convention, it is therefore declared that the identified systematic influence, the drug's effect, is responsible for any difference in mean blood pressure changes between the drug treatment group and the placebo treatment group.

3.3 Statistically Significant Efficacy and Hypothesis Testing

In all clinical research it is important to start with a useful research question, which can be defined by two characteristics: it needs to be specific and to be testable. A general research question such as 'Is this drug good for people's blood pressure?' is not useful. A more useful research question is: Does the drug lower blood pressure statistically significantly more than placebo? Once this research question has been formulated, two hypotheses are created, the research hypothesis and the null hypothesis. (The strict form of the research question should use the word "change" rather than "lower," since we should allow for the possible, even if vanishingly unlikely result that the new drug that is intended to lower blood pressure will actually *increase* blood pressure more than placebo. However, for present discussions, we will simplify the question to make the desired points).

The research hypothesis (sometimes called the alternative hypothesis) typically reflects what is 'hoped for,' which in this case is as follows: The drug lowers blood pressure statistically significantly more than the placebo. In strict scientific terms, such hope has no place in experimental research: the only legitimate hope is to discover the truth, whatever it may be [2]. In the real world, however, this ideologically pure stance is not common for many reasons (financial reasons being not the least of them). The null hypothesis is the counterpart of the research hypothesis. In this case, the null hypothesis is as follows: The drug does not lower blood pressure statistically significantly more than the placebo. There are therefore three items that facilitate hypothesis testing:

1. The research question: Does the drug lower blood pressure statistically significantly more than placebo?
2. The research hypothesis: The drug lowers blood pressure statistically significantly more than placebo.
3. The null hypothesis: The drug does not lower blood pressure statistically significantly more than placebo.

Once the data have been collected in the trial the appropriate statistical test will provide one of two answers. One possible answer says that mean decrease in blood pressure in the drug treatment group is statistically significantly different from the mean decrease in blood pressure in the placebo treatment group, allowing the statement to be made that this difference is unlikely to be the result of chance alone. The second possible answer says that the treatment group means do not differ statistically significantly; this answer can be given even though the group means will almost certainly differ somewhat. This answer allows the statement to be made that this difference could well have arisen by chance alone.

In more formal statistical language, hypothesis testing revolves around two actions following an appropriate statistical analysis:

- Rejecting the null hypothesis;
- Failing to reject the null hypothesis.

Statistical methodology necessitates a choice being made here: it is a forced choice paradigm. It is always the case that one, and only one, of these two actions will occur.

3.3.1 A Straightforward Example

Consider these hypothetical data from a randomized, placebo-controlled, parallel-groups Phase III clinical trial of a new antihypertensive drug. The total number of subjects used in this example is just 21, an extremely unrealistic number considering that an actual trial of this kind may have around 3000 subjects, half receiving the drug and half receiving the placebo. Nonetheless, this very small dataset can be used to illustrate the key points. The following hypothetical data represent decreases in MAP from baseline to the end of the treatment period:

- Drug treatment group ($N = 10$): 3, 0, 3, 8, 5, 9, 4, 7, 5, 6. Mean group decrease $= 5$
- Placebo treatment group ($N = 11$): 4, 2, 2, 0, 1, 0, 1, 4, 3, 2, 3. Mean group decrease $= 2$

Two statistical analyses can be used to analyze these data. First, consider the independent-groups t-test. The term independent-groups reflects the fact that the two treatment groups comprise different individuals (independent-groups analyses are used for data from parallel-groups studies). Like many inferential statistical analyses used to evaluate efficacy, this t-test results in two values. The first is called a test statistic, and the second is the associated probability value or p-value. The name t-test is used since the test statistic provided by this test is called t.

It is not necessary to present all of the calculations necessary when conducting the independent-groups t-test on these data, since this book focuses on conceptual presentations rather than computational presentations. Accordingly, we will focus on the results of the analysis and the interpretation of these results. The following test statistic results from the analysis:

$$t(19) = 3.27$$

3.3.2 A Quick Detour: Degrees of Freedom

The number 19 associated with the test statistic of 3.27 represents the associated degrees of freedom. In this case, the number 19 results since it is two less than the total of subjects in both groups, i.e., 21. Many times when results are presented in CSRs and publications in medical journals the degrees of freedom are omitted, but nonetheless they are vital components of the methodology used to determine the final result from the statistical test, and are therefore worthy of explanation.

Consider the following instruction: Select any five numbers that add up to 100. How much choice is there in this selection? A few moments of thought will reveal that only four numbers can be chosen freely. Once these have been chosen, the fifth number is determined by the four already chosen: Whatever the sum of the first four numbers, addition of the fifth must produce the number 100. If the four numbers already chosen add up to <100, you will need to add a positive number to reach 100. If those already chosen exceed 100 (e.g., 120), you will need to add a negative number (-20) to move back down to 100. Therefore, in this scenario there are four degrees of freedom to your choice; four choices have freedom, the fifth does not.

Now consider the following instruction: Select any five numbers that have a mean of 20. While worded differently, this is an equivalent instruction since any five numbers that add up to 100 will have a mean of 20. Again, therefore, only four numbers can be chosen freely. Generalizing this concept in the context of present discussions, in a particular treatment group of size N with a given mean, there are $(N - 1)$ degrees of freedom. In any study, the number of subjects in a given treatment group is known, and the mean of the measurement of interest (in this case the mean decrease in MAP) can be precisely calculated, and will indeed be calculated shortly. In the present context, the degrees of freedom associated with each group's mean decrease in MAP are $(N - 1)$.

Extending this logic, there are two treatment groups of interest in our current example, and the number of subjects in each group is known. Let us refer to the number of subjects in these groups as N(treatment) and N(placebo). In each case, the number of degrees of freedom is $(N - 1)$. The total number of degrees of freedom in the independent-groups t-test is calculated as the sum of the degrees of freedom for each treatment group:

$$[N(\text{treatment}) - 1] + [N(\text{placebo}) - 1]$$

This can be expressed more succinctly as:

$$N(\text{treatment}) + N(\text{placebo}) - 2$$

The statement "The test statistic obtained at the end of a statistical analysis determines whether or not the result of the statistical test achieves statistical significance" is a correct statement, but an incomplete one. The complete statement is that the test statistic obtained, in conjunction with the associated degrees of freedom, determines whether or not the result of the statistical test achieves statistical significance. The same test statistic may attain statistical significance in conjunction with one set of degrees of freedom and not attain statistical significance with another.

The degrees of freedom associated with a given test statistic are often not cited in regulatory documents or in clinical communications, and so you may not see these very much in your own reading. Nonetheless, it is important to be aware that the level of statistical significance attained by a given test statistic is governed by the degrees of freedom associated with the test statistic.

3.3.3   Returning to Our Example

At the end of Sect. 3.3.1, we had reached the following point in the analysis of our hypothetical data:

$$t(19) = 3.27$$

The next step is to determine the p-value associated with this test statistic and its associated degrees of freedom. In this case, the p-value is 0.004. Hence, the following numerical result can be stated:

$$t(19) = 3.27, p = 0.004$$

This, however, is not the end of the analytical process. These numbers must be interpreted in words in the context of the specific study. We can make the following statements:

1. Since 0.004 is <0.05, the test statistic has attained statistical significance at the 0.05 level. We can therefore reject the null hypothesis, and the result can therefore be declared statistically significant. At this point, the test statistic, i.e., 3.27, can be disregarded: it has determined whether or not statistical significance was attained (it was), and the work of the test statistic itself is therefore done.
2. We now know that the mean decreases in MAP in the two treatment groups are statistically significantly different. To find the direction of the difference we need to calculate and compare the two means.
3. The mean MAP decrease for the drug treatment group was 5 mmHg, and the mean MAP decrease for the placebo treatment group was 2 mmHg. Therefore, the drug led to a statistically significantly greater decrease in MAP than did the placebo.

   We can now calculate the treatment effect. In the present case, the treatment effect is calculated as the mean decrease in MAP for the drug treatment group minus the mean decrease in MAP for the placebo treatment group. Hence, the treatment effect is $5 - 2$ mmHg, i.e., 3 mmHg. We can now give a complete answer: The drug lowered blood pressure statistically significantly more than placebo, and its treatment effect was 3 mmHg. This answer provides compelling evidence of statistically significant efficacy.

3.3.4 A Second Method of Analysis: ANOVA

We noted in Sect. 3.3.1 that two statistical analyses can be used to analyze these data. We have considered the independent-groups t-test. The other appropriate analysis is the one-factor independent-groups analysis of variance (ANOVA). The term independent-groups in this case is derived in exactly the same way as it was for the independent-groups t-test, i.e., independent-groups of subjects were employed in the trial whose results are being analyzed. The term one-factor relates to the fact that, in our ongoing example, there is only one factor that is of interest, i.e., the treatment administered, either the drug or a placebo. A factor is an influence that one wishes to study. In this case, it is of interest to know whether the single factor, the drug being administered, is a systematic source of influence on, and therefore a systematic source of variance in, the data collected in a study. The test statistic in ANOVAs is called F, and the test is typically called the F-test. The name pays respect to Sir Ronald Fisher, who developed this approach.

A very reasonable question at this time is: Since the t-test is perfectly adequate for analyzing data from this kind of clinical trial, what advantage—if any—does ANOVA bring to the table? The answer comes in two parts: It does not bring any advantage in the specific case of our example, in which we have only one factor of interest and only two treatment groups. However, various ANOVAs, which are simply extensions of this simple ANOVA, are capable of analyzing data from many other experimental designs.

The next point to consider is this: Since both the independent-groups t-test and the one-factor independent-groups ANOVA are capable of (and appropriate for) analyzing our hypothetical dataset, they must provide the same answer. Indeed they do, but two points should be noted. First, the formats of the results are different. Unlike the t-test, in which the test statistic t is associated with a single value representing the degrees of freedom, the test statistic F is associated with two values representing the degrees of freedom. Second, the values of the test statistics are not the same: the value of F is the square of the value of t. If this ANOVA were to be conducted on these data, then, the result would be presented as:

$$F(1, 19) = 10.69, \quad p = 0.004$$

Given that the p-value given by the ANOVA test is identical to the p-value given by the t-test ($p = 0.004$), exactly the same degree of compelling evidence of a statistically significant difference in responses to the drug and the placebo is provided.

3.4 Clinically Significant Efficacy and Confidence Intervals

The last section provided evidence of statistically significant efficacy for the drug. However (and it is a big "However"), the attainment of statistical significance, while one of the important factors taken into account by regulatory agencies when considering whether or not to grant marketing approval, is not the whole story. It is certainly correct that, other factors being equal, the greater the treatment effect the more likely it is to attain statistical significance. However, in scenarios where the within-groups variance is particularly low and the size of the study is particularly large it is certainly possible for a small treatment effect to be shown to be a statistically significant effect. Demonstration of a statistically significant treatment effect, therefore, is not by itself informative about the clinical significance of the treatment effect. Gardner and Altman [3] commented that "presenting p-values alone can lead to them being given more merit than they deserve. In particular, there is a tendency to equate statistical significance with medical importance or biological relevance." It is not appropriate to equate statistical significance with medical importance or biological relevance: the latter must be evaluated separately. Just as we saw in the previous chapter when evaluating degrees of risk, the use of confidence intervals is a meaningful strategy in clinical research.

The treatment effect obtained in our ongoing example is 3.00 mmHg (the decimal places have been added since the confidence intervals about to be presented are calculated to two decimal places). It is now appropriate to place a confidence interval around this value, which, as noted earlier, is now referred to as the treatment effect point estimate. In this situation, the two-sided 95% CI is commonly used to determine a range of values that we are 95% confident will cover the true but unknown population treatment effect. The result for these data is:

$$95\% \text{ CI } = 3.00\,(1.08,\ 4.92)$$

The following statement can now be made:

- The data obtained from this single trial are compatible with a treatment effect in the general population as small as 1.08 mmHg and as large as 4.92 mmHg, and our best estimate is 3.00 mmHg.

Using round numbers for the lower and upper confidence interval limits we can say that, if the drug were to be approved and used in the general population, the data from this single clinical trial suggest that it could lower MAP somewhere in the range of 1–5 mmHg. The question now becomes: Has this result provided compelling evidence of clinically significant efficacy or not?

For any set of data, statistical significance can be evaluated by following the procedural rules of hypothesis testing, a precise formulaic strategy that provides an unequivocal answer that a result either is or is not statistically significant. The process of determining clinical significance, however, is not as straightforward since it is not formulaic and requires skilled clinical judgment. Consider our example involving MAP. In clinical practice, antihypertensive therapy is largely, and successfully, based around certain milestones that represent delineation between normal blood pressure and elevated blood pressure. The practicalities of such large-scale pharmacotherapy are assisted by the development of drugs that have worthwhile efficacy, since all drugs bring with them the potential for side effects: the benefit (efficacy) must outweigh any risk. How, therefore, is 'worthwhile' operationalized? This is a matter of clinical judgment.

Framing this question in another way is helpful: What is the smallest treatment effect that is clinically meaningful, or clinically relevant, and hence can be considered clinically significant? This treatment effect size can be called the clinically relevant difference (CRD). Its determination is a clinical one, not a statistical one. Certainly, this determination will be strongly influenced by existing empirical evidence, but its determination is not simply formulaic. Consider the result from our hypothetical example: If the drug were to be approved and used in the general population, the data from this single clinical trial suggest that it could lower blood pressure somewhere in the range of 1–5 mmHg in the general population of hypertensives. Looking at the top end of this range, this question can be asked: Is it clinically relevant, i.e., clinically significant, to decrease blood pressure by 5 mmHg? The answer may be yes. (*Note*: The author is not a clinician, and my comments concerning these hypothetical data should be judged in that light.) At the other end of the range of values, the same question can be asked for a decrease in blood pressure of 1 mmHg. Even though theoretically any decrease is to be welcomed, from a practical pharmacotherapy perspective, a drug that may lower MAP by 1 mmHg may not be considered a good candidate for marketing approval since there are likely to be other drugs already on the market that lower it to a greater degree. Hence, the answer to the question "Is a decrease of 1 mmHg clinically significant?" may be no. Therefore, it is possible that a decision would be reached that the data provided by this trial do not provide compelling evidence of clinically significant efficacy, even though they do provide compelling evidence of statistically significant efficacy.

While this may initially sound like a contradiction, it actually provides a good example of the difference between statistical significance and clinical significance. Think of it in these terms. Statistical significance addresses the reliability of the treatment effect: we have compelling evidence that the drug will lower MAP. In contrast, clinical significance addresses the magnitude of the treatment effect. Putting these together, the data from this single trial provide compelling evidence that the drug will lower MAP in the general population of hypertensives, but it is possible that it will reliably lower MAP by only 1 mmHg.

3.5 Emphasizing an Earlier Point

A point made earlier is worth repeating here. The ultimate purpose of the results from a clinical trial is not to tell us precisely what happened in that trial, but to gain insight into likely drug responses in patients who would be prescribed the drug should it be approved for marketing. This is a fundamental concept in assessments of both safety and efficacy. Appropriate analyses of the data from a given trial, i.e., from a given set of subjects who participated in the trial, are used to infer what is likely to be the case for patients in the future.

References

1. Turner JR (2011) A concise guide to clinical trials. Turner Medical Communications LLC, Chapel Hill
2. Turner JR (2010) New drug development: An introduction to clinical trials, 2nd edn. Springer, New York
3. Gardner MJ, Altman DG (1986) Estimation rather than hypothesis testing: confidence intervals rather than p-values. In: Gardner MJ, Altman DG (eds) Statistics with confidence. British Medical Association, London

Chapter 4
Confidence Intervals: Additional Commentary

Abstract Focus in this chapter falls on two-sided confidence intervals. A two-sided confidence interval constitutes a range of values that are defined by the lower limit and the upper limit of the interval placed around the treatment effect calculated from the trial, which is now referred to as the treatment effect point estimate. The confidence interval is a range of values that is likely to cover the true but unknown population treatment effect with a specified degree of certainty. Three commonly used levels are the 90, 95, and 99% confidence levels. This chapter explains that, when considering the information provided by confidence intervals in their review process, regulatory agencies focus on the limit that represents the 'worst case scenario.' For efficacy considerations, the lower limit represents the least likely benefit. For safety considerations, the upper limit represents the greatest risk.

Keywords Confidence levels · Confidence intervals · Null value of zero · Null value of unity · Width of confidence interval

4.1 Introduction

The usefulness of confidence intervals in analyzing both safety and efficacy data has been noted in the two previous chapters. This chapter provides additional commentary on this topic to emphasize the importance of their employment in the analysis of data collected during clinical trials. Some of the text recaps and extends points made already, and other text introduces new information.

4.2 The Logic of Confidence Intervals

When a randomized later phase clinical trial is conducted, a sample of subjects is selected from the population of all possible subjects, and each member of this sample is placed at random into the drug treatment group or the control drug

J. R. Turner, *Key Statistical Concepts in Clinical Trials for Pharma*,   33
SpringerBriefs in Pharmaceutical Science & Drug Development,
DOI: 10.1007/978-1-4614-1662-3_4, © The Author(s) 2012

treatment group. Analysis of the trial's data provides a precise result for that particular sample. However, while the sample can contain several thousand subjects, this is quite likely to be a (very) small percentage of the population from which that sample was chosen, i.e., the population of individuals in the general population who might be prescribed the drug in the future should it be approved. The primary goal of the trial is not to observe what responses were shown by the subjects who participated, but to use the results obtained to gain insight into what would likely be seen in the population of potential future patients.

If a different sample of subjects had been chosen, the chances of the data obtained being identical is so infinitesimally small that we can say that this would not happen. Therefore, we would have obtained a different dataset. The question of interest here is: How different would the data likely be? Ideally, we would like them to be similar to those obtained in the original trial, thus generating a result that is similar to the result of the original trial: The more similar the results from a second trial, the more confidence we could reasonably place in the results from the original trial.

The word confidence in the previous sentence occurs in the general language use of the word. In the discipline of Statistics the word is used in a specific way when discussing confidence intervals. While one-sided confidence intervals are legitimately used in certain circumstances, present discussions focus on two-sided confidence intervals. A two-sided confidence interval constitutes a range of values that are defined by the lower limit and the upper limit of the interval. These limits are placed around the treatment effect calculated from the trial, which is now referred to as the treatment effect point estimate. The confidence interval thus created is a range of values that is likely to cover the true but unknown population treatment effect with a certain degree of certainty. While this degree can theoretically take any percentage value greater than zero and less than one hundred, three commonly used levels are the 90, 95, and 99% confidence levels.

4.3 Differing Confidence Intervals Differ in Width

Placing a confidence interval around a treatment effect point estimate allows the following statement to be made, where "XX%" can be replaced by 90, 95, and 99%:

- The two-sided XX% CI placed around the treatment effect point estimate obtained from this single clinical trial (or from this meta-analysis) is a range of values that we are XX% confident will cover the true but unknown population treatment effect.

The widths of the three confidence intervals (90% CI, 95% CI, and 99% CI) will be different from each other:

- The 90% CI will be the narrowest;
- The 95% CI will be wider than the 90% CI and narrower than the 99% CI;
- The 99% CI will be the widest.

To move from having 90% confidence that the range of values between the lower and upper limits of the confidence interval covers the true but unknown population treatment effect to having 95% confidence that it does so, the range of values must be greater. That is, we must encompass a wider (greater) range of values to have greater confidence. The same logic applies when moving from having 95% confidence to having 99% confidence that the range of values between the lower and upper limits of the confidence interval covers the true but unknown population treatment effect.

4.4   Employment of Confidence Intervals in Both Safety and Efficacy Analyses

We have seen in Chaps. 2 and 3, respectively, that confidence intervals are informative in the analysis of both safety and efficacy data. In this chapter, additional commentary is provided, and a 'compare and contrast' approach is taken to discuss certain aspects of their nature and use. First, we will note that the term 'treatment effect' is used in both the safety and the efficacy contexts. In the former, the treatment effect captures an unwanted effect of the drug, and in the latter it captures the desired effect of the drug. Ideally, the 'safety treatment effect' is as small as possible, and, when compared with the therapeutic benefit of the drug, it is an 'acceptable' adverse effect. In the case of efficacy assessment, the 'efficacy treatment effect' is ideally large enough to be considered to provide clinically significant therapeutic benefit, and to outweigh any safety concerns, thus leading to a favorable benefit-risk balance or profile.

4.4.1 Equidistant and Non-Equidistant Confidence Intervals

The lower and upper limits of the confidence interval placed around the treatment effect point estimate calculated in the first safety analysis discussed in Chap. 2, i.e., the TQT study evaluating cardiac safety (Sect. 2.8.1), lie equidistantly from the treatment effect point estimate. The treatment effect point estimate is calculated via the mathematical procedure of subtraction, i.e., mean response in the drug treatment group minus mean response in the placebo treatment group. The mathematics of forming the confidence interval involve subtracting a value from the point estimate to create the lower limit, and adding exactly the same value to the point estimate to create the upper limit. Therefore, the lower and upper limits lie equidistantly from the point estimate.

In the second safety analysis discussed, i.e., assessment of the cardiovascular safety of new antidiabetic drugs for T2DM, this is not the case. The reason that this is so is that the treatment effect point estimate is calculated via a different mathematical procedure, i.e., the procedure of division. The subsequent calculations result in lower and upper limits that do not lie equidistantly from the point estimate. In such cases the upper limit will lie further away from the point estimate than will the lower limit.

4.5 The Limit of Primary Interest

In both safety analyses discussed, the upper limit of the confidence interval placed around the treatment effect point estimate was the limit of primary interest. This limit represents the top end of the 'risk scale,' or in other words, how great the risk might be in the population of patients who would receive the drug if approved. In the first case, an examination of drug-induced QT interval prolongation, it represents the greatest likely QT interval prolongation. If this value falls below the threshold of regulatory concern, the regulatory agency will be less concerned than had it fallen at or above the threshold. In the second case, an examination of cardiovascular risk conferred by the drug being developed for T2DM, the upper limit of the relative risk (treatment effect) point estimate represents the greatest likely increase in risk. It is desirable from the FDA regulators' point of view that this upper limit falls below 1.80 initially and, ultimately, below 1.30. In both cases, the lower limit of the confidence interval, which represents the smallest likely risk, is not of immediate interest. The lower limit represents the 'best case scenario,' and regulators must consider the 'worst case scenario' as represented by the upper limit.

In the case of efficacy, it is the lower limit of the confidence interval placed around the treatment effect point estimate that is of primary interest. This limit tells us what the smallest therapeutic benefit of the drug is likely to be in the population of patients who would be prescribed the drug if it were subsequently approved. Therefore, in contrast to the situation for safety analyses, the lower limit represents the 'worst case scenario,' i.e., the smallest likely therapeutic benefit. The upper limit, which represents the best likely therapeutic benefit, is not of immediate interest since regulators must again consider the 'worst case scenario,' here represented by the lower limit.

From the perspective of regulatory decision-making (always intended to protect and promote public health) it is the worst case scenario that is always of primary interest. This is true for both safety and efficacy considerations. In contrast to this similarity, the difference is that the worst case scenario is represented by different parameters in safety and efficacy considerations. In safety considerations, the worst case scenario is the greatest likely degree of risk, captured by the upper limit of the relevant confidence interval. In efficacy considerations it is the smallest likely degree of therapeutic benefit, captured by the lower limit of the relevant confidence interval.

As a final example to illustrate this point in the efficacy realm, consider a randomized, double-blind, placebo-controlled Phase III clinical trial involving a drug intended to raise high density lipoprotein-cholesterol (HDLc), the 'good cholesterol'. One group of subjects receives the drug, and a second group receives the placebo. The magnitude of the drug's treatment effect in this trial (its efficacy) is calculated by subtracting the mean magnitude of change in HDLc for subjects in the placebo treatment group from its mean magnitude for subjects in the drug treatment group. Insight into the likely response of the general patient population, should the drug be approved, is provided by two-sided 95% CIs placed around the treatment effect point estimate.

Imagine a point estimate of 8.00 mg/dl, and lower and upper limits of the two-sided 95% CI of 1.50 and 14.50 mg/dl, respectively, i.e., the confidence interval is 8.00 (1.50, 14.50). The following statement can be made:

- The data obtained from this single trial are compatible with a treatment effect in the general population as small as 1.50 mg/dl and as large as 14.50 mg/dl, and our best estimate is 8.00 mg/dl.

The first point to consider is: Is the value represented by the treatment effect point estimate clinically valuable? (The author is not a clinician, and comments here regarding these hypothetical data should be regarded in this light.) Let us say that, yes, an increase in HDLc of 8.00 mg/dl is clinically valuable, or, in other words, clinically significant. The next question, however, concerns the 'worst possible scenario' as represented by the lower limit of the confidence interval. As noted, the data from this trial are compatible with a treatment effect in the patient population of as small as 1.50 mg/dl. This next question, therefore, is: Is an increase in HDLc of 1.50 mg/dl clinically valuable? While all increases are theoretically beneficial, it might be decided that this increase is too small to warrant moving forward to drug approval, especially if there are already drugs on the market with equally good (or better) safety profiles and that confer greater therapeutic benefit. This may be the outcome even though the upper confidence limit (the best case scenario) provides evidence that the therapeutic benefit may be as high as 14.50 mg/dl.

4.6 Relationship Between Confidence Intervals and Probability Levels

We have seen that the 5% probability level is used in the process of hypothesis testing, which translates into the p-level of $p = 0.05$. If the p-level attained by the test statistic that results from the statistical analysis being performed is less than 0.05, i.e., $p < 0.05$, the null hypothesis is rejected and the result is declared statistically significant. We have also discussed how confidence intervals yield information concerning the clinical significance of the result. It is now appropriate to note that confidence intervals also provide some information on statistical significance, although it is not as detailed as the information provided by the formal inferential test of statistical significance.

In reality, in the situations we have discussed, we would always conduct the formal (inferential) statistical test, e.g., *t*-test, ANOVA, first, and we would not typically proceed to calculate confidence intervals if the inferential test did not generate a statistically significant finding. Therefore, by the time we calculate confidence intervals that, in addition to providing information concerning clinical significance, provide some information on statistical significance, we would already know this information. Nonetheless, the ability to obtain information on statistical significance from confidence intervals can be useful if you see results in published reports of clinical trials that, for some reason, do not include the results of inferential analysis but do provide confidence intervals.

4.6.1 Cases in Which the Null Value is Zero

Confidence intervals are intimately related to probability levels in that levels of statistical significance can be deduced from the values of the confidence interval limits. In situations in which the treatment effect point estimate about which the confidence interval has been placed was calculated via the mathematical process of subtraction, e.g., the efficacy of a drug being developed to lower blood pressure or increase HDLc, the following statement can be made:

- If the two-sided 95% CI that is placed around the treatment effect point estimate excludes zero, the treatment effect calculated in the trial attained statistical significance at the $p < 0.05$ level.

If the confidence interval excludes zero, one of two scenarios must occur: both the lower and the upper limit lie above zero; or they both lie below zero. The latter case would mean that the drug performs statistically significantly *worse* than placebo, an unlikely but nonetheless theoretically possible scenario, and so we will focus on the former. What does it mean when the 95% confidence interval excludes zero? It means that, at the 95% level of confidence, we can state that zero is not a plausible value for the true but unknown population treatment effect.

This relationship between confidence intervals and probability levels arises from the fact that, in the scenario of interest here—the possible difference in response for two treatment groups—the null value is zero. That is, the treatment effect size that would result if the mean responses for each group were identical is zero. The name 'null value' relates to the fact that, if the ultimate form of the null hypothesis that is used in hypothesis testing in this context were true (i.e., there is no difference at all between the treatment groups), the numerical difference between the mean responses for the two groups would be zero. Therefore, if the 95% confidence interval excludes zero, there is compelling evidence at the 5% level that the effect size is not zero, which means that there is compelling evidence that the treatment group means do indeed differ.

Confidence intervals, therefore, can be used to deduce information about statistical significance. However, while they can show whether or not a given level of statistical significance was attained (e.g., $p < 0.05$), they do not yield precise p-values. That is, p-values calculated from an inferential statistical test, e.g., of 0.04 and 0.02, could not be differentiated from a confidence interval analysis alone. Even though 0.02 is considerably smaller than 0.04 (and hence indicates a greater degree of statistical significance attained than does 0.04), both would simply be indicated as $p < 0.05$ by the confidence interval analysis. In regulatory submissions and clinical communications it is much more preferable to present precise p-value resulting from the inferential analysis (which as noted, would in reality be conducted before the calculation of confidence intervals), and then to present the confidence interval analysis subsequently.

The same logic holds true for other levels of confidence. Consider the 99% level of confidence. The following statement can be made:

- If the two-sided 99% CI that is placed around the treatment effect point estimate excludes zero, the treatment effect calculated in the trial attained statistical significance at the $p < 0.01$ level.

The following was noted in Sect. 3.2:

A probability of 5% can be expressed as $p = 0.05$. The statement $p < 0.05$ (i.e., a probability that is less than 0.05) means that the odds of something occurring is less than one in 20, i.e., less than 5%. It has become statistical convention that, if the p-value that results from an inferential test is less than 5%, the result is declared to be statistically significant.

We can now make similar statements for the $p < 0.01$ level of statistical significance. A probability of 1% can be expressed as $p = 0.01$. The statement $p < 0.01$ (i.e., a probability that is less than 0.01) means that the odds of something occurring is less than one in 100, i.e., less than 1%. The $p < 0.01$ is a higher level of statistical significance than the $p < 0.05$ level: That is, attainment of the $p < 0.01$ level of statistical significance provides even more compelling evidence than attainment of the $p < 0.05$ level. (As noted, attaining $p < 0.05$ level itself is considered sufficient to declare a result to be statistically significant in the circumstances discussed in this book).

4.6.2 Cases in Which the Null Value is Unity

The null value as discussed in the previous section is not zero in all circumstances. In other cases the null value is unity, represented as 1.00. In fact, we have already discussed cases where this is true. In situations in which the treatment effect point estimate about which a confidence interval is placed was calculated via the mathematical process of division, e.g., assessing the cardiovascular relative risk of a drug being developed for T2DM as compared with a placebo, the following statement can be made:

- If the two-sided 95% CI that is placed around the treatment effect point estimate excludes unity, the treatment effect calculated in that trial attained statistical significance at the $p < 0.05$ level.

If the number of cardiovascular events occurring in the drug treatment group were exactly the same as the number occurring in the placebo group, the calculation of the relative risk ratio would yield a value of precisely 1.00. Therefore, in this case, the null value (representing absolutely no difference between the groups) is 1.00.

If the confidence interval excludes unity, one of two scenarios must occur: both the lower and the upper limit lie above 1.00; or they both lie below 1.00. The latter case would mean that the drug performs statistically significantly *better* than placebo: It not only does not confer an unacceptable cardiovascular risk, it actually confers a cardiovascular benefit. That is, statistically significantly less cardiovascular events were seen in the drug treatment group than in the placebo group. This is an unlikely scenario for most drugs, but in the case of drugs being developed for cardiovascular disease this is exactly what is desired, and, in that context, having both the lower and upper limit of the confidence interval fall below unity would be compelling evidence of the drug's efficacy.

Returning to our discussion of cardiovascular relative risk for all non-cardiovascular drugs, what does it mean when the 95% confidence interval excludes unity? It means that, at the 95% level of confidence, we can state that unity is not a plausible value for the true but unknown population treatment effect. When both the lower limit and the upper limit fall above 1.00, there is evidence at the $p < 0.05$ level that the drug confers statistically significant cardiovascular risk.

As for the case discussed in the previous section, i.e., when the null value is zero, the same logic holds true for the 99% level of confidence:

- If the two-sided 99% CI that is placed around the treatment effect point estimate excludes unity, the treatment effect calculated in that trial attained statistical significance at the $p < 0.01$ level.

Chapter 5
Meta-Methodology

Abstract Meta-methodology facilitates the quantitative evaluation of the evidence provided by two or more individual clinical trials that have addressed the same research question. It commonly involves not only the statistical combination of summary statistics from various trials (study-level data), but also refers to analyses performed on the combination of subject-level data. Reasons for employing this methodology include: providing a more precise estimate of the overall treatment effect of interest (in the efficacy or the safety realm); evaluating an additional efficacy or safety effect that requires more power than any of the individual trials incorporated could provide; evaluating an effect in a subgroup of participants, or a rare adverse event in all participants; and assessing the possibility of a systematic effect among apparently conflicting study results.

Keywords Meta-analysis · Meta-methodology · Homogeneity · Robustness · Fixed-effect model · Random-effect model

5.1 Introduction

The term meta-analysis is used to describe the quantitative evaluation of the evidence provided by two or more individual trials that have addressed the same research question. It commonly involves not only the statistical combination of summary statistics from various trials (study-level data), but it also refers to analyses performed on the combination of subject-level data. Meta-analyses have attained increasing prominence in the evidence-based medicine literature. The technique has both strengths and weaknesses, and both advocates and detractors. Turner and Durham [1] commented as follows:

J. R. Turner, *Key Statistical Concepts in Clinical Trials for Pharma*,   41
SpringerBriefs in Pharmaceutical Science & Drug Development,
DOI: 10.1007/978-1-4614-1662-3_5, © The Author(s) 2012

If all of the components involved in conducting a meta-analysis are performed appropriately, and the extent to which the results are helpful is not overstated (that is, any limitations are appropriately acknowledged and shared whenever and wherever communicating the results), the results can be informative and instructive. Unfortunately, however, it is easier that one might suspect to conduct a meta-analysis inappropriately and then to overstate the results in a variety of circumstances.

A key reason why it is easier than one might suspect to conduct a meta-analysis inappropriately is that conducting a meta-analysis involves much more than correctly performing some mathematical calculations. Certainly, an analysis is conducted, but the procedure is also critically dependent on choosing the appropriate analytical strategy, interpreting the numerical results of the analysis in the context of the research question being asked, and consistently presenting the results with scientific and clinical decorum. Therefore, as noted by Kay [2], "to ensure that a meta-analysis is scientifically valid, it is necessary to plan and conduct the analysis in an appropriate way. It is not sufficient to retrospectively go to a bunch of studies that you like the look of and stick them together!" Even though conducting a meta-analysis does not require a new trial to be conducted, it is still a research method in its own right.

Given this acknowledgment, the term meta-analysis, while at first appearing appropriately descriptive, does not adequately capture the need for "methodological rigor" [3] in the full array of required actions. Turner [4] therefore suggested that the term meta-methodology can be meaningfully employed to convey the need for paramount methodological rigor when conducting the full array of actions required for its meaningful execution. Perhaps both terms can be meaningfully used, with meta-analysis referring to the computational and interpretive process involved in executing the central analysis itself, and with the term meta-methodology being an all-encompassing name that refers to all necessary meta-methodological aspects of conducting this form of investigation, i.e., the preparation of the dataset to be analyzed, all appropriate analyses that must be conducted (including tests for heterogeneity and robustness), and the appropriate presentation of results and interpretations in all venues and circumstances.

5.2 Unbridled Bravado is Inappropriate When Presenting Results

Once meta-analysts (researchers who have conducted a meta-analysis) have published a report of their work in a journal, some of them unfortunately disseminate their findings in the mass media with a bravado that markedly departs from calm, scientific, and clinical discourse, and seemingly with the expectation that the nation's physicians will change their practice of medicine immediately [4]. As Turner et al. [5] noted, "In the era of sensationalist, sound-bite coverage, clinical science sadly falls very low on the list of points to be covered in the allotted 30 seconds of television coverage," a point not unknown to meta-analysts who willingly participate in this circus. Fortunately, many more authors take a more circumspect approach, providing physicians and their patients with well-reasoned and appropriately presented benefits and potential harms of a given intervention in the spirit that treatment decisions be made, as they should be, by physicians and their patients on a case-by-case basis. The intervention is made when physician–patient agreement is reached that the benefit-risk balance is favorable (benefit-risk balances are discussed in more detail in the following chapter).

As is true across all research methodology, if the correct study design has been employed and rigorous methodology has permitted the acquisition of optimum quality data, the computational analysis itself is typically not difficult. The results are therefore not difficult to calculate. The maturity and intellectual honesty of the meta-analysts lie in the interpretation of the results and the appropriate degree of restraint needed to disseminate their conclusions in a responsible manner. Given all of these considerations, meta-methodology must be undertaken carefully, diligently, and responsibly.

5.3 Fundamentals of Meta-Methodology

In the case of an analysis of results from a collection of randomized clinical trials the methodological rigor required by each individual trial is also required for all aspects of facilitating, conducting, and reporting the new analysis. Just as each trial had a Study Protocol and a Statistical Analysis Plan (or a statistical section in the Protocol) written before its commencement, a similar approach is required here too. The term meta-methodology captures this requirement well. As in experimental methodology, the optimum quality answer to the research question of interest is dependent upon the employment of optimum quality study methodology and appropriate statistical analysis that is dependent upon the nature of the data themselves [4].

Meta-methodology facilitates a quantitative evaluation of the evidence provided by two or more individual trials that have addressed the same research question. It commonly involves not only the statistical combination of summary statistics from various trials (study-level data), but it also refers to analyses performed on the combination of subject-level data. Reasons for the employment of meta-methodology include:

- Providing a more precise estimate of the overall treatment effect of interest (the treatment effect can be in the efficacy or the safety realm);
- Evaluating whether overall (positive) results are also seen in prespecified subgroups of participants;
- Evaluating an additional efficacy or safety effect that requires more power than any of the individual trials incorporated can provide;
- Evaluating an effect in a subgroup of participants, or a rare adverse event in all participants;

- Assessing the possibility of a systematic effect among apparently conflicting study results.

The conceptual basis of meta-methodology is straightforward: more data provide a better opportunity to get an optimum-quality answer to a research question. However, as alluded to previously, appropriate selection of studies to be included, implementation of the appropriate statistical techniques, and the appropriate interpretation and communication of the results obtained are not so straightforward.

The purpose for a meta-methodological study must be stated at the outset, and the treatment effect of interest determined and identified [4]. Subsequently, the basic steps are:

- Establishing rules for which of the studies about to be identified will be incorporated;
- Identification of all pertinent studies;
- Data abstraction and acquisition;
- Data analysis;
- Evaluating heterogeneity;
- Evaluating robustness;
- Dissemination of results and conclusions.

One straightforward approach is to include every study identified and obtained. A counterargument is that, almost certainly, some studies will be 'better' than others, and that 'less good' studies should perhaps not be included if optimum quality information is to be provided to the clinicians who will read publications of the results and therefore may base treatment decisions on these results. In the latter case, a priori inclusion and exclusion criteria that operationally define 'better' and 'less good' must be stated in advance of searching for studies, thereby determining which of the studies about to be identified will be included in the analysis. Identification of all studies which may potentially be included has become much easier with the advent of computer search engines and web-based tools, but it can still be a challenging task, particularly when trying to locate and obtain unpublished data (publication bias is a powerful potential confounding factor in this context).

Data extraction and acquisition also has its challenges. Many published studies of individual trials present summary statistics as opposed to presenting the underlying data, i.e., the subject-level data. While Piantadosi [6] noted that "more than the published information is usually necessary to perform rigorous overviews," this requires the meta-analysts to obtain and analyze subject-level data (not simply aggregate data across treatment groups) from each study. Unless a sponsor is conducting an analysis of data from several of its own studies, as can now be the case during the development of new anti-diabetic drugs for T2DM as discussed in Chap. 2, it can be difficult to gain access to subject-level data for every study of interest. For this reason, meta-analyses based on summary measures of the treatment effect gathered from published studies are currently more commonly reported. Accordingly, discussions in this chapter pertain to such analyses.

5.4 Data Analysis: Fixed-Effect and Random-Effect Models

The goal of meta-methodology is to estimate the overall treatment effect of interest across all of the studies included, i.e., to obtain a single estimate of the treatment effect and the variance associated with it. Two items of data are obtained from each study incorporated:

- A measure of the treatment effect in that specific study;
- An estimate of the variance associated with that specific treatment effect.

However, before conducting the analysis, it must be decided whether to employ a fixed-effect analysis model or a random-effect analysis model. The fundamental difference between the models pertains to the degree of influence each individual treatment effect is allowed to exert mathematically on the overall treatment effect.

Each study included in the analysis contributes the same piece of information to the analysis, i.e., the treatment effect found in that study. Therefore, if the analysis incorporates 100 studies, 100 treatment effects are included. However, each item of information does not necessarily carry the same weight when determining the result of the analysis: some items are accorded more influence than others. This influence is expressed in terms of the weight ascribed to each study and hence to each individual treatment effect. Studies whose treatment effects are weighted more heavily will exert a greater influence on the final result of the analysis than those whose treatment effects are weighted less heavily. The weighting accorded to each study-specific treatment effect is determined computationally according to the rules of the analysis model adopted.

In the fixed-effect model the weight assigned to each study is the inverse of the variance associated with the study's treatment effect. Therefore, the more precise the treatment effect estimate (i.e., the smaller its variance) the greater the study's influence on the overall treatment effect. Larger studies tend to yield more precise estimates, so this weighting system suggests that the overall estimate will be more consistent with those reported in the larger studies incorporated. In the random-effect model the calculation of the weight assigned to each study involves an additional component. The weight assigned to each study is a function of the inverse of the variance of the treatment effect estimate (the sole item used in the fixed-effect meta-analysis model) and an estimate of the between-study variance in underlying parameters, since, unlike the fixed-effect model, the random-effect model acknowledges that these can vary.

This is an intuitively sensible feature of the random-effect model, since in most circumstances it would be very surprising if the studies incorporated did not vary from each other. This variation arises from the fact that each study included in the analysis will likely have many unique features, such as the nature of the study population and the number of subjects in each treatment group within the study, the length of treatment periods, the investigational sites at which the study was conducted, the quality of measurements made, and perhaps even in the precise way the treatment effect of interest was defined and therefore calculated.

While the methods of estimating this additional component need not be discussed here, it is important to recognize the consequences. The more similar the individual studies included in the analysis, the less the impact of this component. That is, the more similar the studies, the closer the results obtained from the two analysis models: if there were no difference between studies the results generated from the two analysis models would be identical. Phrasing this the other way round is perhaps more helpful: The greater the degree of difference between the studies incorporated in the analysis, the more important the choice of analysis model becomes.

The salient advantage of the random-effect model is that it allows between-study variability to influence the overall treatment effect estimate and, more particularly, its precision. This precision is captured by the estimate of its variation, which is represented by confidence intervals placed around the point estimate of the overall treatment effect generated by the analysis method chosen. The random-effect model tends to generate wider confidence intervals, indicating less precision in the point estimate. The consequence of this observation is as follows. If a fixed-effect model (which assumes that the studies included in the analysis do not differ) is employed when there is indeed a considerable difference between the studies included, the confidence placed in the result of the analysis will be greater than it should be. Had a random-effect model (which takes into account differences between the studies) been employed, the confidence interval placed around the point estimate would have been wider, indicating less confidence in the precision of the result obtained.

The relevance of the width of a confidence interval placed around a point estimate is that, for a given point estimate, it is possible that a more narrow confidence interval placed around it would result in a statistically significant finding from the analysis whereas a wider confidence interval would not (recall discussions in Chap. 4). While the clinical significance of the result from such an analysis focuses much more on the confidence intervals placed around the point estimate than on the point estimate itself, achievement of statistical significance alone often drives the media's interest in the result.

5.5 Evaluating Heterogeneity

The statistical theory underpinning such analysis assumes that the study-specific estimates of the treatment effect are (relatively) homogenous. Homogeneity is present when the study-specific estimates are similar in magnitude and direction to the estimate of the treatment effect resulting from the combined analysis. Heterogeneity can arise from differences between studies, such as the possibilities already noted. Since the objective is to calculate a well-justified combined estimate of the treatment effect of interest, a formal evaluation of homogeneity following a visual graphic inspection of the combined effect against each individual effect is a recommended strategy.

This formal evaluation involves a statistical test, such as the Cochran Q test. Homogeneity (also expressed as lack of heterogeneity) is indicated by a statistically non-significant result. While general acceptance of the $\alpha = 0.05$ criterion provides a 'line in the sand' that is useful in certain circumstances, blind adherence to characterizing a result as either statistically significant or not statistically significant using the $\alpha = 0.05$ level is not necessarily a clinically meaningful strategy. In this context, a statistically non-significant Cochran test can be (mis)interpreted to state that there is no heterogeneity present. That is, a fallacious argument can be made that the lack of statistically significant evidence of heterogeneity represents an all-or-none statement of its complete absence [7].

5.6 Evaluating Robustness

Having calculated the result of the analysis it can be informative to assess its robustness. In any combined analysis, some of the studies included will be larger than others, and sometimes a small percentage of included studies can be considerably larger than the majority of others. As already noted, the nature of the calculations performed here means that the larger trials tend to influence the result more, since they tend to have greater precision.

It can therefore be helpful to assess the robustness of the overall conclusion by performing the analysis without the data from the largest study or studies to see if the results remain qualitatively the same. If they do, then the result of the primary overall analysis is deemed robust. If they do not, confidence in the overall result can be undermined. Moreover, if the results are considerably different, it simply may not be appropriate to present the combined result alone and make statements based on it.

While the word 'qualitative' may initially seem surprising here, qualitative assessments can complement primary quantitative assessments in various instances. Another example of qualitatively different results obtained with various analytical approaches leading to a lesser degree of confidence in the interpretation of results from a study is seen in the analysis of efficacy data presented in pre-approval clinical trials. Data from two analysis populations, the Intent-to-Treat (ITT) and the Per-Protocol populations, are typically used in two separate sets of efficacy analysis. Regulatory agencies are generally much more encouraged if the two sets of results are qualitatively similar (the probability of them being quantitatively identical is vanishingly small). If they are not, questions may be raised as to why they are not. Additionally, if the Per-Protocol population is a lot smaller than the ITT population (it will almost certainly be somewhat smaller), regulatory reviewers will wonder why. Were there a lot of major protocol violations? Were a lot of subjects removed for the same protocol violation? Were many of the subjects with protocol violations enrolled at the same investigative site? Are there any systematic problems in the conduct of the trial? All of these issues can reduce overall confidence in the trial's findings [8].

5.7 Hypothesis Generation and Hypothesis Testing

The discipline of statistics, like the practice of medicine, can be described as a science and an art. Performing numerical calculations is the easy part: knowing how best to design a study to answer a research question, collecting the appropriate data, and then analyzing them using the appropriate statistical procedures is the artful and skillful part. Among many skillful actions is strict adherence to the statistical convention dictating that a set of data generating a research hypothesis cannot be used to test that hypothesis. As one example of where this convention is important, consider subgroup analysis conducted using data from a large clinical trial. The response of well-defined subgroups of subjects to a therapeutic drug is a topic of legitimate medical interest. It may well be biologically plausible that, on average, members of certain well-defined subgroups would respond differently than members of other well-defined subgroups or from the study population as a whole, meaning that efficacy and/or safety concerns may be quite different for them. Therefore, it is clinically legitimate to address this question.

Investigation of such differences may be stated a priori as a secondary endpoint analysis in a trial's protocol or statistical analysis plan. It is also possible that a biologically plausible subgroup difference that had not been anticipated may be identified in post hoc analysis, thereby generating a research hypothesis for subsequent testing. The key point is that such findings are best thought of as 'hypothesis generating' (creating a question) and not as hypothesis testing (answering a question). To answer a question arising in this matter it is appropriate to conduct a subsequent trial where the item of interest is the primary endpoint analysis, with the study powered appropriately to answer the question (test the hypothesis). With regard to meta-methodology, if a research question is generated by inspection of a given data set from a given trial, those data should not be included in a combined analysis data set used to test the hypothesis generated: to do so is to operate in the methodological twilight zone.

Going a step further, a result from a meta-analysis is best thought of as hypothesis generating, i.e., creating a question, than hypothesis testing, i.e., answering a question [3].

5.8 Results and Conclusions

The results of a completed analysis would typically include the following:

- The treatment effect point estimate for each individual study included in the analysis, and a confidence interval (often the 95% interval) about each study's estimate;
- The overall treatment effect point estimate and its confidence interval.

This information can be displayed in tabular form, or in a graphical form called a confidence interval plot.

As noted earlier, the definition of the term meta-methodology includes the appropriate presentation of results and interpretations in all venues and circumstances, i.e., presenting them in a calm manner with scientific and clinical decorum.

References

1. Turner JR, Durham TA (2009) Integrated cardiac safety: assessment methodologies for noncardiac drugs in discovery, development, and postmarketing surveillance. Wiley, Hoboken
2. Kay R (2007) Statistical thinking for non-statisticians in drug regulation. Wiley, Chichester
3. Kaul S (2011) Cardiovascular safety in drug development: state-of-the-art assessment. Presentation given at the DIA/FDA/CSRC-sponsored meeting, April 14. Washington DC
4. Turner JR (2011) Editor's Commentary: Additional Associate Editors, New Submission Category, and Meta-methodology. Drug Inform J 45:221–227
5. Turner JR, Satin LZ, Callahan T, Litwin J (2010) The science of cardiac safety. Appl Clin Trials 19(Suppl 11):4–8, 11
6. Piantadosi S (2005) Clinical trials: a methodologic perspective, 2nd edn. Wiley-Interscience, Hoboken
7. Diamond GA, Bax L, Kaul S (2007) Uncertain effects of rosiglitazone on the risk for myocardial infarction and cardiovascular death. Ann Intern Med 147:578–581
8. Turner JR (2010) New drug development: an introduction to clinical trials, 2nd edn. Springer, New York

Chapter 6
Benefit–Risk Estimation

Abstract Benefit–risk estimation is a key facet of decision-making at both the regulatory (public health) level and the level of individual patient pharmacotherapy. It requires consideration of both benefit and risk. For a drug to receive regulatory approval, the regulatory agency must find a drug to have a favorable benefit–risk profile: The benefit to the population as a whole must outweigh any potential risk to certain individuals, who will be protected by appropriate drug labeling and other risk-mitigation strategies. At the individual patient level, a prescribing physician and a patient must believe that the potential benefit to the patient outweighs any potential risk. Benefit–risk estimation is therefore a key component of Integrated Pharmaceutical Medicine.

Keywords Benefit–risk estimation · Benefit–risk balance · Decision-making · Quantitative benefit–risk estimation · Threshold of regulatory concern

6.1 Introduction

Benefit–risk estimation requires consideration of both benefit and risk. The term benefit–risk estimation (rather than the commonly seen term benefit–risk assessment) is deliberately used here since its estimation requires the comparison of two items which themselves are estimates, i.e., estimate of benefit and estimate of risk. Hence, the term assessment can carry an implication of precision that is not the case.

Benefit is the easier of the two to address. Risk is sometimes referred to as harm. From the perspective of the media's treatment of drug-related issues, the use of the term harm (while clinically meaningful) may not be an optimal choice since it would bring added emotional fuel to reporting that already often contains way too much.

J. R. Turner, *Key Statistical Concepts in Clinical Trials for Pharma*,
SpringerBriefs in Pharmaceutical Science & Drug Development,
DOI: 10.1007/978-1-4614-1662-3_6, © The Author(s) 2012

The term drug safety falls lower on the emotional thermometer, but defining safety is more difficult than might initially be thought. This chapter therefore starts with a discussion of how best to do so before progressing to consider benefit–risk balances.

Decision making by a sponsor is a critical consideration throughout drug development. At the end of several phases of development (and sometimes at the end of a single trial) a decision to progress or not progress to the next stage of development, typically called a go/no-go decision, must be made. Should the drug reach the point where a marketing application is submitted to a regulatory agency, the agency will need to decide whether or not to approve the drug, and, if it is to be marketed, whether any stipulations should be made regarding the drug's use. Following that, patients and their physicians will consider whether a treatment is suitable on a case-by-case. Benefit–risk estimations provide useful information for making all of these types of decisions. Since the discipline of Statistics facilitates the estimation of benefit and of risk, it provides the rational basis for informed decision-making.

At the regulatory (public health) level, a new drug must have a favorable benefit–risk balance to receive marketing approval. Second, at the level of the individual patient, the prescribing physician and the patient must decide that a particular course of pharmacotherapy has a favorable benefit–risk balance. In both cases benefit–risk estimation can be represented as follows [1]:

$$\text{Benefit-risk estimate} = \frac{\text{Estimate (probability and degree) of benefit}}{\text{Estimate (probability and degree) of harm}}$$

Consider benefit first, which is operationalized in clinical trials as efficacy and, once the drug has been marketed, as effectiveness. The inclusion of both probability and degree addresses the fact that the likelihood the drug will work is just as important as the degree to which it works when it does work. The same is true for risk, which captures both the probability and degree characteristics of harm. If the probability of a relatively serious unwanted drug response is 1 in 1,000,000, the estimation of the benefit–risk balance for a given estimate of benefit is quite different than if the probability of the side effect is just 1 in 100.

6.2 Drug Safety

Drug safety was introduced in Sect. 1.5.1. As noted there, a useful definition of drug safety in terms of benefit–risk estimation was provided by the FDA's Sentinel Initiative [2]:

> Although marketed medical products are required by federal law to be safe for their intended use, *safety does not mean zero risk*. A safe product is one that has acceptable risks, given the magnitude of benefit expected in a specific population and within the context of alternatives available.

In addition to considering safety per se, this representation makes clear that benefit must be considered as well.

6.3 Decision Making

If forced to summarize the purposes of study design, experimental methodology, operational execution, and statistical analysis in one sentence each, the following might be suitable [1]:

- *Study design*: Designing a clinical trial to facilitate the collection of data, i.e., unbiased and precise numerical representations of biologically important information that best answers the study's research question.
- *Experimental methodology*: Considering and implementing all necessary procedures that, if executed correctly, allow the acquisition of optimum quality data.
- *Operational Execution*: Conducting all operational and experimental tasks correctly and therefore actually acquiring optimum quality data.
- *Statistical analysis*: Describing, summarizing, analyzing, and interpreting the data collected to answer the study's research question.

Expanding on the last point, numerical representations of biologically important information facilitate answers to questions that arise during the process of new drug development and thus provide the basis for making the best possible decision at each time given the best evidence available at that time. (It is quite appropriate to use additional information to come to a new decision at a later time).

Many decisions during the process of new drug development concern whether or not to proceed to the next step in the process. Adequate evidence needs to be obtained, and documented, to permit careful consideration of the benefits and risks of proceeding. Given a finite amount of resources and, particularly in the case of larger pharmaceutical companies, a choice of drug candidates upon which to focus these resources, it is financially prudent to proceed only if there is a reasonable chance of success (the definition of 'reasonable' being unique to each sponsor and drug candidate). While business driven, the choice to pursue development programs that are likely to yield successful drugs is arguably in the best interests of patients. Pursuing development plans for drug candidates that are likely to fail reduces the sponsor's ability to work on drugs that may get approved and help patients.

6.4 The Subjective Nature of Many Decisions

It is perhaps intuitive to think that science produces clear-cut answers, and scientists pride themselves on the objectivity inherent in their disciplines. Clinical science, clinical research, and clinical practice, however, require a combination of objective information and informed judgment. Since all judgment is subjective, subjectivity is an integral part of clinical science, clinical research, and clinical practice.

In this context, the word subjective does not carry the potentially negative connotations that may accompany it in other realms. All of us would likely welcome the medical opinion of an experienced and well-informed clinician when making a decision concerning several possible therapeutic regimens. The opinion offered would be the clinician's best clinical judgment based on the best available evidence at that time. In the context of study design, Piantadosi [3] made the following comment:

> It is a mistake not to recognize the subjectivity that is present, or to design and interpret studies in formulaic ways. We could more appropriately view experimental designs as devices that encapsulate both objective plans along with unavoidable subjectivity (p.131).

The discipline of Statistics certainly involves using informed judgments. Statistics is an art as well as a science, a sentiment expressed well by Katz [4]:

> No degree of evidence will fully chart the expanse of idiosyncrasy in human health and disease. Thus, to work skillfully with evidence is to acknowledge its limits. All of the art and all of the science of medicine depends on how artfully and scientifically we as practitioners reach our decision. The art of clinical decision making is judgment, an even more difficult concept to grapple with than evidence (p. xi).

Consider also the decisions that must be made by regulatory agencies. From many perspectives, the role of regulatory agencies is far from easy. For example, they have to decide if it is appropriate to allow a sponsor to commence clinical testing (Phase I trials) based on data submitted in an Investigational New Drug Application (IND). No animal model is a perfect predictor of human responses to an investigational drug, and so the decision to allow a sponsor to commence clinical trials requires a judgment call. An enormous amount of information has to be provided to regulatory agencies to allow them to make this decision, but it is still a judgment call, albeit a well-informed one.

The same is true when a regulatory agency evaluates the evidence presented in a New Drug Application (NDA) for a small-molecule drug or a Biologics License Application (BLA) for a biologic. Again, a tremendous amount of information is presented following the conduct of the clinical trials that comprise the drug's preapproval clinical development program, but these data cannot guarantee that rare serious adverse drug reactions will not be seen once the drug is approved and taken by a large number of patients. The agency has to evaluate all of the data in the marketing application and consider the benefit–risk ratio of approving the drug, an unenviable responsibility. While the randomized controlled trials that we have discussed in this book remain the gold standard for evaluating the efficacy of a new drug in preapproval clinical trials, and they do provide (some) safety data, they cannot be regarded as the sole source of safety data or as a guaranteed predictor of the drug's effectiveness in the target population. As the Institute of Medicine noted, "The approval decision does not represent a singular moment of clarity about the risks and benefits associated with a drug—preapproval clinical trials do not obviate continuing formal evaluation after approval" [5]. In a real sense, marketing approval of a new drug can be regarded as the beginning of its true evaluation.

6.5 Determining and Enforcing Thresholds of Regulatory Concern

When employing a model to prospectively exclude unacceptable risk, as discussed in Chap. 2, we need to combine clinical science, regulatory science, and statistical science. Regulatory scientists must consider the clinical evidence and then determine the thresholds to be employed. Choosing these thresholds may be a better expression than determining them: there is no precise formulaic manner to facilitate determination. Therefore, there is a degree of subjectivity in the creation of a regulatory threshold too.

At this point we can expand on our considerations of one particular threshold of regulatory concern. In Sect. 2.8.1 we discussed the Thorough QT (TQT) study, and noted that the threshold of regulatory concern had been set at 10 ms. That is, a mean QT interval prolongation of 10 ms or more would lead to regulatory concern. We then considered a scenario in which the treatment effect point estimate for QT interval prolongation in a TQT trial was 8.00 ms, and the lower and upper limits of a two-sided 90% CI placed around this point estimate were 6.50 and 9.50 ms, respectively. This result can be written as 8.00 (6.50, 9.50). We noted that the following statement concerning the true but unknown treatment effect in the general population from which the sample that participated in the trial was drawn could be made:

- The data from this single TQT study are compatible with a treatment effect (prolongation of the QT interval) as small as 6.50 ms and as large as 9.50 ms in the general population, and our best estimate is a treatment effect of 8.00 ms.

The upper limit value of 9.50 ms falls below this threshold of 10 ms, and therefore this particular drug would not be deemed to be associated with unacceptable cardiac risk.

However, the situation becomes more challenging when the upper limit falls above but fairly close to the threshold. Consider a second scenario in which the treatment effect point estimate for QT interval prolongation in a QT trial was also 8.00 ms, but the lower and upper limits of a two-sided 90% CI placed around this point estimate were 5.00 ms and 11.00 ms, respectively, a result that would be written as 8.00 (5.00, 11.0). The following statement would be made:

- The data from this single TQT study are compatible with a treatment effect (prolongation of the QT interval) as small as 5.00 ms and as large as 11.00 ms in the general population, and our best estimate is a treatment effect of 8.00 ms.

In this scenario, consideration of the upper limit in a strict, black-and-white sense would lead one to the conclusion that the drug is indeed of regulatory concern, and therefore potentially not suitable for marketing because of its cardiac safety risk. The ICH E14 guideline itself uses the terms negative study (upper limit of the two-sided 90% CI is less than 10 ms) and positive study (upper limit of the two-sided 90% CI is 10 ms or greater). These terms suggest an inflexible 'approve/do not approve' dichotomy that is actually not the case.

The true purpose of the TQT study is not to facilitate a decision at that time that, no matter how good all other data look, the drug will not be approved should the threshold of regulatory concern be breached. Rather, it is to determine the degree of QT interval (and general ECG) monitoring that must be conducted in subsequent Phase III trials. The considerably larger number of subjects that will participate in these later trials will provide the opportunity for a more extensive evaluation of the degree of cardiac risk. The greater the degree of QT prolongation observed in the TQT study, the greater the regulatory concern, and therefore the greater the likelihood that the sponsor will be expected to conduct more intensive and extensive cardiac monitoring in later clinical trials than would typically be conducted for a drug in the same class. However, it is also the case that the greater the severity of the drug's indication, and the fewer available treatments that are already available on the market, the more likely a drug is to be approved for a given degree of QT prolongation. Benefit–risk estimation is therefore a crucial aspect of a regulatory agency's deliberations when considering granting marketing approval. If approved, the drug's labeling will carry information concerning its degree of QT prolongation.

6.6 Decision Analysis and Decision Making

Benefit–risk estimations are not themselves the final step in the process. Rather, they provide the most rational foundation we have for making decisions. These decisions include sponsors' go/no-go decisions during clinical development programs, regulatory marketing decisions, regulatory postmarketing decisions to modify a drug's labeling or availability, and patient–doctor treatment decisions.

During each stage in a drug's development and use, new data concerning the benefits and risks of the drug can emerge. In early stages of development, estimates of benefits (i.e., efficacy in terms of a clinically relevant endpoint) are associated with a great deal of uncertainty owing to the relatively small sample sizes of early phase studies. This inherent uncertainty regarding the drug's benefit–risk profile can complicate the decision-making process of sponsors who must make a number of decisions regarding further development choices among multiple drug candidates within their portfolio. These decisions are often made in the broader context of the competitive marketplace by considering the emerging product profiles of other drugs in development. With a number of competing products on the market, the product with the greatest advantage would be the one with the most favorable benefit–risk profile in this use of the term, as evidenced by its labeling. Certain drugs may well have identified 'risks' made clear in their approved labeling.

Many sponsors use decision analysis to prioritize the products in their portfolios. Decision analysis enables the sponsor to quantify various aspects of a decision so that the consequences of various decision alternatives are more apparent. Comparisons of alternatives are made possible by construction of a decision model, which identifies the various decisions to be made (e.g., initiate a Phase III trial, conduct an additional Phase II trial, or abandon development of that particular drug) and the consequences of each decision.

Decision making must continue throughout a marketed drug's entire life on the market. It is appropriate to re-evaluate previous decisions on the basis of new information, and reaching a different decision than was reached in the past is not a statement (an admission) that the previous decision was 'wrong.' Rather, it is a statement that additional information now suggests a different decision. We can only make decisions with the information available at the time of the decision. Decision making is predicated on benefit–risk estimates that require calculation.

6.7 Qualitative and Quantitative Considerations

While the qualitative concept of the benefit–risk balance (and hence of a favorable benefit–risk balance) is readily apparent, our quantitative abilities in this arena are not well developed. Garattini [6] observed that "The benefit–risk balance itself has no generally recognized measure," one of the reasons why the FDA Sentinel Initiative regards benefit–risk analysis to be "one of the important facets of the science of safety that urgently requires additional development" [2]. Turner [7] discussed this issue, providing references to several papers addressing this topic. See also the article by Tominaga et al. [8] and the references it provides.

References

1. Turner JR (2010) New drug development: an introduction to clinical trials, 2nd edn. Springer, New York
2. FDA (2008) The Sentinel Initiative: National Strategy for Monitoring Medical Product Safety (see www.fda.gov)
3. Piantadosi S (2005) Clinical trials: a methodologic perspective, 2nd edn. Wiley, Hoboken
4. Katz DL (2001) Clinical epidemiology and evidence-based medicine: fundamental principles of clinical reasoning and research. Sage Publications, Thousand Oaks
5. Institute of Medicine (2007) The future of drug safety: promoting and protecting the health of the public. National Academies Press, Washington

6. Garattini S (2010) Evaluation of benefit-risk. Pharmacoeconomics 28:981–986
7. Turner JR (2011) Editor's commentary: letters to the editor, a new publishing partner, and benefit-risk estimation. Drug Inform J 45:559–564
8. Tominaga T, Asahina Y, Uyama Y, Kondo T (2011) Regulatory science as a bridge between science and society. Clin Pharmacol Ther 90:29–31

About the Author

J. Rick Turner, Ph.D., is an experimental research scientist and clinical trialist, and currently Senior Scientific Director, Integrated Cardiovascular Safety, Quintiles. With his colleagues, he provides Sponsors with consultation, strategic and regulatory insights, and operational support during cardiac safety assessments throughout clinical development programs. He is particularly interested in the development of drugs for Type 2 Diabetes Mellitus and obesity.

Dr. Turner is also a Senior Fellow at the Center for Medicine in the Public Interest (New York) and Editor-in-Chief of the DIA's peer-reviewed *Drug Information Journal*. He has published 12 previous books, 65 peer-reviewed papers, and many articles in professional journals. His books include:

- Turner JR, 2012, *A Concise Guide to Clinical Trials*. Chapel Hill, NC: Turner Medical Communications LLC.
- Turner JR, 2010, *New Drug Development: An Introduction to Clinical Trials*, 2nd Edition. New York: Springer.
- Turner JR, Durham TA, 2009, *Integrated Cardiac Safety: Assessment methodologies for noncardiac drugs in discovery, development, and postmarketing surveillance*. Hoboken, NJ: John Wiley & Sons.
- Durham TA, Turner JR, 2008, *Introduction to Statistics in Pharmaceutical Clinical Trials*. London: Pharmaceutical Press.

Index

J. R. Turner, *Key Statistical Concepts in Clinical Trials for Pharma*,
SpringerBriefs in Pharmaceutical Science & Drug Development,
DOI: 10.1007/978-1-4614-1662-3, © The Author(s) 2012